HEAVY METALS: A PROBLEM SOLVED?

ENVIRONMENT & POLICY

VOLUME 22

Heavy Metals: A Problem Solved?

Methods and Models to Evaluate Policy Strategies for Heavy Metals

Edited by

Ester van der Voet,
Jeroen B. Guinée
and
Helias A. Udo de Haes

Centre of Environmental Science,
Leiden University, Leiden, The Netherlands

Authors:

Jeroen C. J. M. van den Bergh, Mathijs N. Bouman and Patricia P. A. A. H. Kandelaars
Faculty of Economics,
Vrije Universiteit, Amsterdam, The Netherlands

Theo M. Lexmond, Simon W. Moolenaar
Department of Soil Science and Plant Nutrition,
Wageningen Agricultural University, Wageningen, The Netherlands

Jos Boelens, Xander Olsthoorn
Institute for Environmental Studies,
Vrije Universiteit, Amsterdam, The Netherlands

Evert Verkuijlen
Interfaculty Department Environmental Science
University of Amsterdam, Amsterdam, The Netherlands

Mathijs N. Bouman, Jeroen B. Guinée, Reinout Heijungs, Gjalt Huppes,
Lauran van Oers, Helias A. Udo de Haes, Ester van der Voet
Centre of Environmental Science,
Leiden University, Leiden, The Netherlands

SPRINGER-SCIENCE+BUSINESS MEDIA, B.V.

A C.I.P. Catalogue record for this book is available from the Library of Congress.

ISBN 978-90-481-5406-7 ISBN 978-94-015-9610-7 (eBook)
DOI 10.1007/978-94-015-9610-7

Printed on acid-free paper

Originally published by Kluwer Academic Publishers in 2000
Softcover reprint of the hardcover 1st edition 2000

Table of Contents

Abstract

Environmental problems related to heavy metals have a long history. Certain events in the past have induced governments to address these problems in their environmental policy. As a result, the emissions from point sources have been reduced significantly over the past decades in many countries. Some therefore may regard the metals' problem to be solved. However, the inflow of freshly mined metals into the world economy has remained at a high level. The question then is, if the metals no longer are emitted, what then is their fate? This question was the motive for the start of an interdisciplinary research program, the Metals program, financed by the Dutch National Science Foundation (NWO). This research program had two aspects: (1) development of methods and models to address such problems, and (2) by applying these methods and models reaching conclusions on the nature of the societal metabolism of four heavy metals in the Netherlands, the risks involved and the possibilities for a sustainable metals management: copper, zinc, lead and cadmium. The main conclusions from the Metals program can also be grouped according to those two aspects.

Methods and models
Integrated, quantitative modelling of the flows and accumulations of metals such as cadmium, zinc, copper and lead, together with their driving forces and their environmental risks, has led to substantial new insights in these metals' metabolism and in relevant management options. In dealing with the complexity at hand, we have found that an overview of the relevant information can be obtained rather by successive use of different models aiming at the answering of different questions, than by the development of one encompassing integrated model. Examples are the subsequent use of Substance Flow Analysis and Environmental Risk Assessment, and Substance Flow Analysis and Materials-Product Chain analysis. For the evaluation of the outcome of the various developed models, a set of sustainability indicators was developed. A clear and explicit definition of such indicators, both related to parameters in the economy as well as in the environment, has proven to be very useful as a basis for environmental policy as well as for scientific development.
Another conclusion was that a clear picture of the risks connected with the metals' metabolism can only be obtained if the modelling is performed at different levels of scale. Starting from a national level, at a higher scale level attention can be given to the analysis of problem shifting by pollution export, whereas at a lower level more detail can be obtained about actual risks in specific situations, such as different agricultural practices.

Metals management
The past reduction of emissions of the investigated metals to water and air appears to have gone in hand with an increase of the accumulation of these metals in stocks of materials, products and landfilled waste. As a result, a future rise in emissions to the environment will take place if no action is taken. On the long run, this will lead to a surpassing of water and soil standards for ecosystem health and of standards for human exposure through the different environmental media. Thus the present day use of the

four metals cannot be regarded as sustainable. In the surpassing of thresholds a key role is played by so-called trace flows, in contrast to the bulk flows which are generally well managed. Management measures should therefore focus on the control of these trace flows, rather than on a still further enhancement of recycling. In agriculture specific risks occur due to the occurrence of "closed loop accumulation", being accumulation in fodder-soil-fodder cycles of particularly copper and zinc. On short notice, the following measures appear to be feasible: reduction of zinc and copper concentration in fodder, terminating the use of metal based pesticides, and prevention of corrosion by coating or a partial substitution of applications in the built environment. Even with the implementation of the above measures, in non-disruptive policy scenarios political threshold values are expected to be surpassed in the long run. Other measures will be required, especially the immobilisation of metals in solid waste flows and a complete phase-out of many applications, including non-functional ones. Although there is ample time for the implementation of such measures, the question is whether such a strong sustainability approach is feasible at all in view of the many other environmental problems which must be addressed. If not, an adjustment of the present standards in the direction of weak sustainability may well become unavoidable.

Preface

This publication contains the results of an interdisciplinary research programme, the 'Metals' programme, financed by the Dutch National Science Foundation (NWO). This was one of the three research project clusters of NWO's Sustainability and Environmental Quality Programme, the aim of which was to find ways of operationalising the policy-related concepts of 'sustainability' and 'environmental quality' in a scientifically sound manner. As the title suggests, the Metals programme focused on the issue of heavy metals. The core research problem is the fact that although metals emissions have declined considerably over the last few decades, mining of these substances has remained at more or less the same level. The main research questions studied in the programme concern the fate of the mined metals, whether this fate is in any way connected with environmental risks and, if so, how to render the metals management regime more sustainable. To answer these questions it was necessary to operationalise the concepts of 'sustainability' and 'environmental quality' from the perspective of a society's management of heavy metals.

In the Metals programme these questions were addressed by an interdisciplinary team of ecologists, agricultural scientists, environmental scientists and economists. The approach adopted in the research programme was based on development and application of economy-environment models. Given the scope of the research programme, many different issues have been examined. Many issues remain unresolved, however, and many new questions have arisen in the course of work. Consequently, the outcomes of the models developed do not provide any definite answers. Nevertheless, the general direction of the results is sufficiently robust for a number of conclusions to be drawn regarding the present metals management regime as well as the basic direction in which it should be changed. Insight was gained, moreover, in the specific difficulties regarding integrated economy-environment modelling. Last but not least, we experienced what it means to cooperate in an interdisciplinary team, which was not easy in the beginning but proved to be very rewarding in the end.

Part I Introduction

Contents:

In Part I we introduce the research problem: the actual and potential environmental consequences of the production and use of heavy metals. We argue that, contrary to general expectations, the metals problem might not one the decline. The five main research questions are stated, basic concepts are outlined and the methods of investigation employed to seek answers to these questions are presented.

I.1 General Introduction

Ester van der Voet, Helias A. Udo de Haes & Jeroen B. Guinée

I.1.1 Environmental problems related to the use of materials

Human society is facing many problems related to the environment. A significant number of these problems are the consequence of current modes of processing materials and energy. On the one hand, there are the problems related to resource depletion. These were first signalled several decades ago (Club of Rome, 1972) and have since lost some of their urgency: geological stocks are not as scarce as first believed and substitution of one stock by another may take place, moreover. Today the emerging opinion is that the main depletion problem in fact concerns biotic renewable resources. On the other hand, the processing of materials and energy leads to pollution problems. By dispersing substances in the environment natural processes are disturbed with a number of potentially adverse consequences, including direct threats to human health, ecosystem damage and economic damage, through a wide variety of mechanisms.

Large-scale impacts are caused by the human addition to biogeochemical cycles of C, N, P, S, water and other substances, transforming these into anthropo-biogeochemical cycles. Examples include global warming through the increase of the relatively small atmospheric stock of CO_2 and eutrophication of lakes and coastal waters through the increase of aquatic nutrient stocks. Frequently, it is not directly that enlarged flows cause adverse impacts but indirectly, through the resultant slow increase of relatively small but crucial stocks. Managing these cycles is difficult because these elements are very basic, not only for our economic system but even for maintaining human life: for breathing and feeding. Reducing use of these substances in anything but a marginal way is therefore often not an option; the challenge for management is to bring the human part of these cycles in line with the natural part, either by isolating the human from the natural cycle or by a major shift in the ecological grounding of our society.

On a smaller scale, emissions of micro-pollutants disturb natural processes and pose a threat to human and ecosystem health because of their toxicity, carcinogenicity, mutagenicity or hormone-mimicking properties. Examples include the more persistent organic micro-pollutants and heavy metals, which are emitted in small quantities but accumulate in the environment because of their non-degradability. These substances can likewise be analysed as economic-environmental cycles. In most cases the natural cycle is relatively small and caused only by the weathering of rocks and volcanic eruption. The anthropogenic part of the cycle thus generally predominates. Here, too, environmental stocks are often the key issue: the slow and steady increase of stocks in soils and sediments and the accumulation and bio-concentration in the food chain. Managing these cycles may be easier, on the one hand, for in many cases the human contribution can be reduced by substituting other processes or materials without any major disruption of society. On the other hand, it is more difficult since the influx of even minor amounts to the environment may pose risks and emissions may often elude us because of their low magnitude and their sometimes unexpected occurrence. The challenges involved in managing these cycles are therefore quite different but also substantial.

This book studies the cycles of a number of micro-pollutants and addresses the problems related to heavy metals as they occur in the Netherlands. As in many industrial countries, emissions of these metals to the atmosphere and to surface waters have been reduced considerably over the past few decades. As a result, the emerging opinion in environmental policy circles is that the metals problem has been more or less solved, at least in the Netherlands. Among environmental scientists, however, the nagging feeling remains that this might be too optimistic: although metals emissions have been reduced, mining operations have remained at a high level. This feeling resulted in the combined research programme of which this book is one of the outcomes (Udo de Haes et al., 1992). States very concisely, the purpose of this research programme was to establish whether the environmental problems related to metals have indeed been solved in the Netherlands, and if not, in which direction a sustainable management regime for these metals should be sought.

Below we address the problem of heavy metals in the Netherlands and the reason for doubting the status 'solved' and the research questions with which we set out. We then discuss some of the basic starting points of the research programme.

I.1.2 Environmental problems related to metals

Environmental problems related to heavy metals have a long history. Heavy metals have toxic properties, leading to adverse effects on human and ecosystem health even in small doses. Another problem-causing property is their non-degradability: once they enter the environment they will remain there for a long time. Metals tend to accumulate in soils and sediments, with immobilisation due only to geological, and therefore extremely slow, processes. Accumulation in the food chain may lead to an increased stock in biota, thereby magnifying the human dose.

Well-known examples of metals poisoning in past centuries include the lead poisoning from water pipes in ancient Rome and the mercury poisoning of the 'mad hatters' in Europe (Markham, 1994; O'Carroll et al., 1995). In this century we have seen, among other cases, the tragedy of mercury poisoning in the Minamata Bay in Japan, through consumption of coastal fish, and that of cadmium poisoning through consumption of polluted rice (Japanese Ministry of Health and Welfare, 1968). Lead in petrol has caused health problems in many cities, especially for children (see, for example, Rhode Island Kids Count Factbook, 1997). These and similar incidents have spurred governments to implement environmental policies and industries to reduce their emissions substantially. Comparing current emissions from industrial and other point sources to those of several decades ago, at least in the industrialised countries, there has evidently been a very major reduction (e.g. Ayres & Rod, 1986; Stigliani & Anderberg, 1992). Present policies regarding heavy metals include not only end-of-the-pipe emission reduction but also recycling and even more source-oriented measures limiting or banning certain applications altogether (e.g. Bulletin of Acts, Orders and Decrees of the Kingdom of the Netherlands, 1990). In the Netherlands a comprehensive heavy metals policy is currently being formulated. The general feeling is that the main problems have been solved and that it is now a question of tying up a few loose ends and then continuing to enforce legislation. One of these loose ends is the existence of polluted sites, a relic of the past, described by Stigliani & Salomons (1993) as 'chemical time bombs'. Such sites may become unsuitable for agriculture or housing construction. If they remain unattended metals may become available and leach to the groundwater through increasing soil acidity. Other loose ends refer to applications considered risky, such as metal-based pesticides and paints, but which have not been regulated.

Although emissions in the Netherlands have undoubtedly declined - today the single major source of surface water pollution is the Rhine water entering the country - there are still several matters that are cause for concern. One is the fact that environmental metals concentrations are not decreasing in the Netherlands. This may be due to a time lag - once emissions have been reduced the metals already in the environment disappear only at a geological rate - but it may also have more serious causes. We observe that the inflow of metals into the economic system - equivalent, at a global level, to the amount of metals being mined - has not decreased but has remained at a high level, as shown in Table I.1.1. This global trend also shows up quite clearly in the Netherlands (FOE, 1998).

Table I.1.1: *Global production rates of some metals for the period 1980-1992 (ktonnes/yr.); data cover intentional metal ore production unless noted otherwise.*

	1980[1]	1983[1]	1986[2]	1989[2]	1992[3]
arsenic	25	28	40	40	34
cadmium[4]	18	17	19	21	20
copper	7760	8500	7993	8887	8900
chromium	2717	2500	3625	3737	4025
lead	3186	3240	3335	3395	3200
zinc	6338	6300	6936	7062	7365

[1] USBM, 1985 (Mineral Facts and Problems).
[2] USBM, 1989 (Minerals In The World Economy).
[3] USBM, 1993 (Mineral Commodity Summaries).
[4] Cadmium extracted from zinc ore.

Figures I.1.1 and I.1.2 show the trends of heavy metals emissions to the atmosphere and to surface waters.

Figure I.1.1 Trends in heavy metals emissions to the atmosphere in the Netherlands, 1985 - 1996.

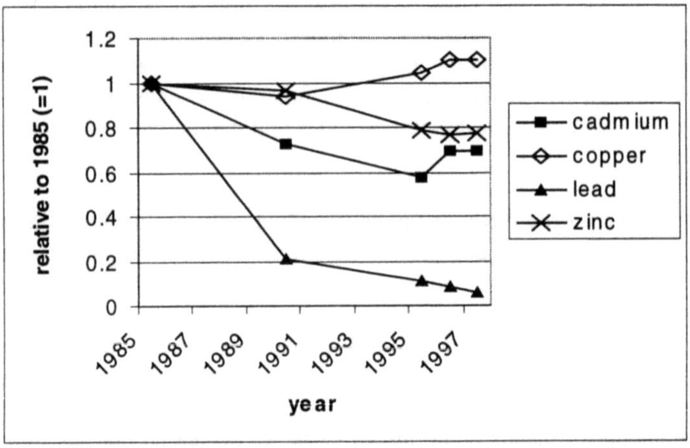

Source: RIVM, 1998.

Figure I.1.2 Trends in heavy metals emissions to surface waters in the Netherlands, 1985 - 1995.

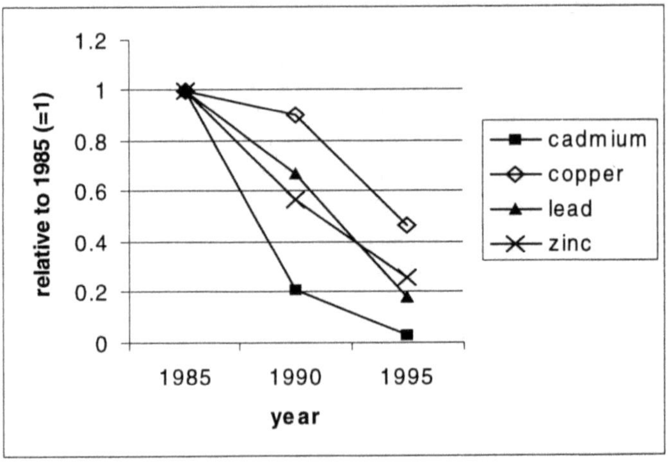

Source: CBS.

Here we quite clearly see the reduction of the emissions over the past 15 years, especially the emissions to water. This raises the question: if emissions have indeed been reduced, then where does the inflow into the economy end up? There are several possibilities in theory:

- Although point-source emissions have decreased we have no insight into the more diffuse emissions. Examples of such emissions include phosphate fertiliser, which is polluted with small amounts of metals and which is emitted directly into agricultural soils. Of such emissions there are no records and they may even have increased.

- Emissions may have been replaced by landfill, i.e. there may have been a shift from emissions to the atmosphere and surface waters to dumping in landfill sites.
- The metals entering the economy may be accumulating in materials and products, thus increasing the societal stock and in due course, i.e. in the waste phase, causing emissions to rise once again.
- The Netherlands may have 'exported' the more polluting stages of the metals' life cycle to other countries, thus enjoying the benefits of consumption while transferring the burden of mining, production and waste management elsewhere.
- Safe storage may have been established for waste metals, reducing emissions from waste materials to zero.

At the outset some of these possibilities already seem more credible than others. We do know, for example, that no storage at present qualifies as safe in the sense of reducing emissions to zero. Pollution export may indeed take place at the level of a small country but this does not explain a decrease in global emissions. The other three possible explanations all appear reasonable. All of them, to varying degrees and in various ways, cause us to query the characterisation of the metals problem as a problem of the past.

I.1.3 Research questions of the Metals programme

The purpose of the research programme has been to establish whether the environmental problems related to metals have indeed been solved in the Netherlands, and if not, in which direction sustainable management of these metals should be sought. The above considerations have led to a number of research questions being addressed in the research programme. These questions are not only scientific but also policy-oriented: in answering them we may arrive at additional recommendations for an environmental policy aimed at metals. The research questions stated in the original application form are the following:

1. *What are the flows and stocks of the selected metals through the economy and the environment?*
   This question can be regarded in the tradition of the concept of 'industrial metabolism', the description of the economy in terms of the processing of materials. This concept is introduced in Section I.2. Having an overview of flows and stocks in society enables one to establish links with environmental flows on the one hand and with economic processes on the other.
2. *How can these flows and stocks be modelled?*
   In order to establish the linkages between economy and environment in a quantitative manner, the aforementioned overview is not sufficient. The relations between flows and stocks and between flows and economic or environmental variables are important from the perspective of metals management. In Section II.1 this will be further elaborated. A number of models have been developed in the course of the research programme; these are described in Part II. The application

of these models to describe and analyse (parts of) the heavy metals problem is treated in Part III.

3. *What is the fate of the mined metals and what are the related environmental risks?* This refers to the inconsistency between the constant level of mining and the sharp reduction of recorded emissions and is in fact the main question. This question is addressed with the aid of an account and a model of metal flows (see Section 1.2).
4. *Is the present metals management regime sustainable?* This question refers to the present situation in the Netherlands but also to future developments, or rather the future consequences of the present management regime, and to the situation in other countries due to pollution export, as mentioned above. For evaluating management in terms of sustainability a number of indicators have been developed. These are treated in Section II.6. In Section I.2 a more general treatment of the concepts of sustainability, environmental quality and sustainable development is presented.
5. *In as far as the present metals management regime is not sustainable, how can we design a management strategy that is?* In the research programme no attempt has been made to draw up a formal method to design a sustainable scenario. In Part IV a comprehensive attempt is made to formulate scenarios 'offhand', based mainly on the results of the analyses of the previous research questions. Three scenarios of increasing stringency, and therefore also of increasing societal disruptiveness, are described and evaluated using the developed models and sustainability indicators.
6. *Can a statement be made with regard to the 'net sustainability' of the Netherlands?* This refers to the possibility of the economies of industrialised countries such as the Netherlands having being 'cleaned up' at the expense of other parts in the world, by locating the more polluting stages of the metals' life cycles such as mining, refinery and waste treatment elsewhere. To signal this a 'pollution footprint' indicator has been developed and applied. This indicator is described in Section II.6.

Some demarcations and methodological choices have been made to focus the research:

- The heavy metals considered are copper (Cu), zinc (Zn), lead (Pb), and cadmium (Cd). The reasons for this choice are both practical (well-investigated metals, therefore good data availability) and theoretical (all four are metals with decreasing emissions and a more or less constant economic inflow). Moreover, all four are addressed by the Dutch heavy metals policy-to-be.
- The geographical boundaries are those of the Netherlands; the territorial waters of the North Sea are not regarded as part of the system.
- Within the Netherlands we have endeavoured to be comprehensive in the investigation of flows and stocks, regarding both the economy and the environment.
- Two economic sectors have been investigated in more detail: the housing sector, because of the large flows and stocks associated with it, and agriculture, where flows are much smaller but involve greater risks to human health.
- The reference year is 1990, for reasons of data availability. For evaluation of the scenarios, the years 2050 and 2100 have been taken. This may seem a rather long time-frame from the perspective of policy formulation, but since the life-span and residence time of metals in both economic and environmental stocks is very long,

the time lag between policy implementation and effects may also be considerable. To establish an adequate basis for comparison, we have also calculated the steady-state situation associated with each of the scenarios.

I.1.4 Multidisciplinary approach

In order to address these research questions as described above, we have taken as our starting point the concept of 'industrial metabolism', the powerful image of the analogy between the processing of matter in the biosphere and the technosphere. For metals, we have attempted to define the normative policy-oriented concepts of 'sustainability' and 'environmental quality' in industrial-metabolic terms. This is treated in more detail in the next section, I.2. The main thrust of the research programme has been to develop models, of both a metabolic/environmental and an economic nature, which can be used to analyse quantitatively the connections between the societal processes of mining, producing and using metals, on the one hand, and environmental problems, on the other. The developed models are intended to support environmental policy by providing answers that cannot be reached using monodisciplinary methods and models: while environmental models may be used to describe the fate of emissions, they do not provide any insight into the economic mechanisms underlying the emissions; and while economic models describe actors' responses, their treatment of the economic consequences is usually less than adequate. Therefore, aspects of both are required to obtain relevant answers to the research questions described above, either within one model or by using a combination of models. This multidisciplinary approach is mirrored in the composition of the group of researchers, which comprised environmental specialists, environmental scientists and economists.

In the course of the research program a number of models have been developed. A description of these models is presented in Part II. The 'core model' is a metabolic model in the tradition of Materials Flow Accounting (MFA) and Substance Flow Analysis (SFA), in line with the core concept of industrial metabolism. MFA and SFA are also treated in Section I.2, as core approaches of this research programme. This model can be used to describe, analyse and predict societal and environmental flows and stocks of metals, thus establishing a quantitative economy-environment relationship in terms of mass flows. More detailed metabolic models have been developed, additionally, for agricultural soils in relation to agricultural management practice. On the economic side, Materials-Product Chain models have been developed to address several aspects of integrated chain management, which is also treated in Section I.2.

Application of the developed models to the case of heavy metals, as undertaken in the course of the research programme, is described in Part III. In Part IV we make a combined effort to apply all the models to one case - development and assessment of three scenarios for heavy metals in the Netherlands - thus looking at one problem from different angles. Part V is devoted to discussion and conclusions.

References

- Ayres, R.U. & S.R. Rod (1986). Patterns and Pollution in the Hudson-Raritan Basin. Environment no. 28, pp 14-20 and 39-43.
- Bulletin of Acts, Orders and Decrees of the Kingdom of the Kingdom of the Netherlands no. 538 (1990). Chemical Substances Act - Cadmium Decree, enacted 12 October 1990.
- CBS (Central Bureau of Statistics): environmental data can be found at http://statline.cbs.nl/witch/selned.htm
- Friends of the Earth (Dutch branch) (1998). Nederland Duurzaam Plus. Report Vereniging Milieudefensie, Amsterdam.
- Japanese Ministry of Health and Welfare (1968). As provided by http://www.kanazawa-med.ac.jp/~pubhealt/cadmium2/itaiitai-e/itai01.html
- Markham, A. (1994). A Brief History of Pollution. Earthscan Publications; Published in association with WWF-UK.
- Meadows, D.H., D.L. Meadows, J. Randers & W.W. Behrens III (1972). Limits to Growth: A Report for the Club of Rome's Project on the Predicament of Mankind. New York: Universe Books.
- O'Carroll, R.E., G. Masterton, G. Dougall & K.P. Ebmeier (1995). The neuropsychiatric sequelae of mercury poisoning; the mad hatters disease revisited. Br J Psychiatry 167(1):95-98.
- Rhode Island Kids Count Factbook 1997, as provided by http://www.rikidscount.org/97fbook/table15.html
- RIVM (National Institute of Public Health & the Environment) (1998). Milieubalans 97, het Nederlands milieu verklaard.
- Stigliani, W. & W. Salomons (1993). Our fathers' toxic sins. New Scientist 11 December 1993, pp 38 - 42.
- Stigliani, W.M. & S. Anderberg (1992). Industrial Metabolism at the Regional Level: the Rhine Basin. IIASA working paper WP-92-10, Laxenburg Austria, 40 pp.
- Udo de Haes, H.A., L. Reijnders, H. Verbruggen, L. Hordijk, J.B. Opschoor, F.A.M. de Haan & Th. G. Drupsteen (1992). Accumulation of metals in economic/environmental cycles: mechanisms, risks and possible management strategies. Granted in the framework of the Research program Sustainability and Environmental Quality, funded by the Dutch Science Foundation.
- United States Department of the Interior - Bureau of Mines (USBM) (1985). Mineral facts and problems. US Government Printing Office, Washington DC.
- United States Department of the Interior - Bureau of Mines (USBM) (1992). Minerals in the World Economy - 1989 International Review. US Government Printing Office, Washington DC.
- United States Department of the Interior - Bureau of Mines (USBM) (1993). Mineral Commodity Summaries. US Government Printing Office, Washington DC.

I.2 Basic concepts and approaches

Ester van der Voet, Helias A. Udo de Haes & Jeroen B. Guinée

I.2.1 Introduction

As argued in Part I.1, substance-related problems - and therefore also those relating to heavy metals - can be regarded in terms of the disturbance of natural biogeochemical cycles. Analysis of the transformation of such cycles into anthropo-biogeochemical cycles requires an integrated economy-environment approach. In this respect it is important to assess human influence on the natural cycles, in order to determine the extent of the anthropogenic disturbance. Furthermore, we need to consider the 'backfiring' mechanisms, i.e. the influence of the artificially enlarged environmental cycles on matters valued by society, such as human health, loss of welfare and well-being, and species and ecosystem health. In this section, a number of concepts are discussed that form the basis of our problem analysis and the quest for solutions presented in Parts II, III and IV. We discuss the following concepts:

- sustainability, environmental quality and sustainable development
- industrial metabolism.

Next, we introduce the main approaches of the 'Metals' programme:

- the Materials/Substance Flow Analysis approach
- the Chain Management/Analysis approach.

I.2.2 Concepts

Sustainability, environmental quality and sustainable development

The concept of *sustainability* was introduced by Brundtland (World Commission on Environment and Development, 1987) as a vision of bridging the gap between economic development and environmental carrying capacity. It has become an important catchword in environmental policy. We discuss it together with the related concepts of sustainable development and environmental quality. Sustainability, environmental quality and sustainable development are heuristic concepts, pointing to fruitful directions of economic behaviour, environmental management and research, rather than well-defined scientific terms. These concepts are used to relate scientific research findings to value judgements: when we hear talk of a 'sustainable' scenario we do not know its contents, but we do know that it is a 'good' scenario. When it comes to specifying the particulars, these concepts are found to be operationalised in many different ways. They must therefore be defined in more detail whenever they are used to evaluate the results of a study. This we endeavour to do below, taking as our starting point the heavy metals issue in the Netherlands.

Environmental quality is related to the state of the environment and changes therein. Environmental quality is, by our definition, sustainable when the various functions required from the environment are not impaired, nor will be in the future. These functions can be divided into a number of categories, such as production functions, regulation functions, carrier functions and information functions (Van der Maarel & Dauvellier, 1978). In addition, the 'intrinsic value' of nature may be included.

Sustainable development on the other hand refers to human activity, and thus to the state of the societal system and changes therein. Human activity (development), again by our definition, is sustainable when it does not impair environmental quality as defined above, now or in the future. Thus, environmental quality and sustainable development are related, and for both, a sustainable level may be defined.

It is not easy to operationalise these concepts, not even within a limited scope such as the metals management of a region. In the first place, the translation from environmental quality to sustainable development and vice versa leaves room for debate since knowledge regarding the various chains of impact may be incomplete. Moreover, such a translation is not merely a technical issue, but depends strongly on choices and societal values. In the second place, the functions required from the environment are not unambiguous themselves. The functions required may change in the course of time, as technology develops and delinkage of the societal and environmental subsystems progresses (Bringezu, 1993; Van der Voet et al., 1996). In the third place, different functions may place different (qualitative) requirements on the environment. It will be difficult to deal with these differences when defining these concepts at a generic level, as is required in the context of environmental policy.

When related to substance flows, the notion of 'environmental quality' is often interpreted in terms of generally accepted concentration limits in the respective environmental compartments. Such an interpretation is a valuable starting point, especially for national policy-makers. In our research programme we also adopted this basic approach: environmental quality is considered sustainable when there is no perceived health risk to humans or ecosystems. In the context of this study we have opted to assess this risk with reference to current policy standards. Thus, there is no perceived risk when standards aimed to protect human and ecosystem health (environmental concentrations and daily intake) are met, now and in the future.

This notion of environmental quality must then be connected to sustainable development. We define a society's metals management regime as sustainable if it does not cause transgression of the aforementioned environmental quality standards, now or in the future. This implies that we must find ways to link society's management of metals to environmental concentrations. We have found a valuable concept in this direction in the shape of *industrial metabolism.* Approaches derived from this concept are *materials/substance flow analysis (MFA/SFA),* which links societal to environmental metabolism, and *integrated chain management*, which takes a more partial view of the societal system and addresses specific economic chains from-cradle-to-grave. Below, the MFA/SFA and the integrated chain management approaches are discussed more extensively.

Industrial metabolism

The concept of industrial metabolism, as defined by Ayres (1989), argues the analogy between the economy and environment on a material level: the economy's 'metabolism' in terms of materials mobilisation, use and excretion to create 'technomass' is compared to the use of materials in the biosphere to create biomass. Whereas in the biosphere processes are attuned to such a degree that waste generated in one process is converted into a resource for another, in the economy resources are squandered, thus creating both depletion and pollution problems. In order to abate and prevent these problems, society must look to the biosphere for guiding principles. The description of the economy thus is limited to a description of the *physical economy*. The research field of *industrial ecology* (Jelinski et al., 1992) is concerned with elaborating and operationalising this concept, and takes the physical economy as its primary object of study.

To regard the economic system in terms of its flows of materials and energy has opened up possibilities for assessing economic development also in environmental terms. This has given rise to many different analytical and applied studies, as well as new directions in spatial design. Analytical studies introduce the study of anthropogenic cycles, which makes it possible to determine the human contribution to environmental flows, to assess the origins of environmental problems, and to detect problem-shifting to other areas or other time periods as a result of changes in the human management regime. They also include the study of cradle-to-grave economic chains, which enables detection of problem-shifting to other substances or environmental problems as a result of there being different ways to fulfil certain functions (one of the first studies in this field was reported by Hunt et al., 1974). Applied studies include, for example, eco-efficiency practices within plant facilities or companies (see, for example, OECD, 1998), or even for society as a whole (von Weiszäcker et al., 1997). In the spatial development domain the so-called industrial ecosystems can be mentioned, aiming at concentrating a number of activities in one area to create scope for designing collective waste treatment processes, re-using and recycling waste materials and energy within the area, etc., according to the biosphere law 'one process' waste is another's resource'. Even outside the materials and energy domain the industrial ecology analogy is fruitful, in the sphere of biotechnology and use of renewable resources, for example.

For the problem of heavy metals as addressed in the 'Metals' programme, we have confined ourselves to the area of materials flows and to the analytical studies. Below, we discuss the approaches we have taken from the industrial metabolism concept in dealing with the case of heavy metals.

I.2.3 Approaches

The Materials/Substance Flow Approach

An important principle in the field of industrial metabolism is the *materials balance*, used as an instrument for describing the materials regime of the economy based on the Law of Mass Conservation, again analogous to the long-standing practice of investigating ecological materials cycles. Materials Flow Analysis (MFA), including Substance Flow Analysis (SFA), is based on this materials balance approach. MFA offers an economic counterpart to the study of ecological materials flows, thus extending the concept of biogeochemical cycles and opening up the potential to study their transformation into anthropo-biogeochemical cycles. There is no generally accepted definition of MFA. According to some, this type of analysis varies widely in scope, encompassing virtually all accounts or models which describe or calculate flows of materials or substances in some way, in the economy or the environment. Others attach a more confined meaning to these tools, limiting them to the consistent investigation of the flows of certain materials or substances in a certain year through a geographically demarcated economy-environment system. We here use the narrower meaning.

The MFA approach includes the analysis of bulk flows (bulk-MFA) as well as the analysis of specific elements or chemical compounds (SFA) (Bringezu et al. (eds.), 1997). Bulk-MFA is used to comment on the materials throughput and the materials intensity of national economies, important sectors or large functional systems and therefore concentrates on bulk or mass flows. It has led to recommendations such as dematerialisation of the economy and de-linking of economic growth and environmental pressure, in its purest form advocated by von Weiszäcker et al. (1997). SFA is used to identify the causes of specific pollution problems in the economy and find ways of resolving or preventing these problems, and is therefore concerned with the flows of specific substances (van der Voet, 1996). Generally speaking, bulk-MFA stops at the 'border' of the environment, while SFA also considers the environmental flows. SFA has been applied in this research programme to the case of heavy metals in its accounting as well as static/steady-state and dynamic modelling application. For this purpose, the SFA model FLUX has been developed during this research programme, as described in Section II.2.

A specific form of SFA is so-called environmental fate modelling. This type of model concentrates on environmental flows. It is based on physico-chemical properties of substances on the one hand and environmental characteristics on the other. Such a fate model can be linked to risk assessment models. In the course of this research program the model Dynabox, described in Section II.5, has been developed. The input for this fate model are the emissions generated by FLUX, and it also contains a risk assessment module.

The approach described above, establishing a quantified link between environmental concentrations and human and ecosystem health risks, on the one hand, and societal metabolism and management, on the other, adds a new element to already existing approaches. Still, such an approach implies a narrowing of the concept of environmental quality. For one thing, locational environmental distribution characteristics are not taken into account. Furthermore, the different demands set by the different functions of the environment are ignored. Sustainable development in this approach is generally translated into emission targets only. This, too, implies a loss of meaning. Issues such as the shifting of problems either to the future or to other countries is also important but are not included in a concentration/emission approach. The view of the world taken from the industrial metabolism concept provides scope for expanding the concentration/emission approach as described above. We did so by defining indicators specifically for the societal management of metals and by including indicators for problem-shifting in space and time, alongside indicators for environmental concentrations and human intake. These indicators are described in Section II.5.

Chain Management and Chain Analysis
Perhaps even more limiting than the issues mentioned above is the fact that by adopting an SFA approach considerations regarding substitution, optimisation and costs drop out of the picture altogether. This not only limits the scope of the evaluation, but it is also problematical from a modelling point of view: in general, economic market mechanisms are the prime mechanisms determining the course and size of flows through society. Here we encounter one of the major difficulties of the research programme as a whole: existing physical models such as SFA models leave out the economic driving forces, while existing economic models ignore physical laws, even if they do have an add-on environmental module, for example in the form of emission coefficients. This discrepancy is treated in Section II.1, where economic and physical modelling principles are discussed.

On the economic side the chain management / chain analysis approach has been used to develop a link between metabolism and economic mechanisms. Chain management and analysis are concerned with economic chains, i.e. chains of connected processes connected with certain economic services, from the cradle (mining of raw materials) to the grave (final disposal of waste materials). Such chains can be analysed in different ways: Life Cycle Assessment (LCA) studies, for example, aim to specify the integrated environmental impacts (Guinée, 1995), while micro-economic models are used to identify means for regulation based on market mechanisms. In the course of this research programme we have tried to merge economic and mass balance considerations in a single modelling approach: the *Material-Product Chain or MPC approach.* MPC analysis tries to integrate physical and economic aspects of material and product flows in a Material-Product Chain (M-P chain), which is defined as a set of linked flows of materials and products fulfilling a certain service (Opschoor, 1994). An analysis of an M-P chain can be defined broadly as an analysis of the structure of connected material and product flows. An economic analysis of an M-P chain focuses on economic aspects like allocation, substitution, recycling and behaviour. Such economic modelling of M-P chains requires combining the elements of physical flow and economic models (Kandelaars, 1998). This approach is described in more detail in Section II.2. In the research programme, MPC analysis has been applied to rain gutters and window frames (Section III.3).

References

- Ayres, R.U. (1989). Industrial Metabolism. In: Ausubel & Sladovich (eds.): Technology and Environment, pp 23-49. Nat. Academy Press, Washington DC.
- Bringezu, S. (1993). Towards increasing resource productivity: how to measure the total material consumption of regional or national economies? Fresenius Envir. Bulletin 2: 437-442.
- Bringezu, S., M. Fischer-Kowalski, R. Kleijn & V. Palm (eds.) (1997). Regional and National Material Flow Accounting: from Paradigm to Practice of Sustainability. Proceedings of the ConAccount workshop 21-23 January 1997, Leiden, the Netherlands. Wuppertal Special no. 4.
- Guinée, J.B. (1995). Development of a methodology for the environmental life-cycle assessment of products, with a case study on margarines. PhD thesis, Leiden University.

- Hunt, R.G., W.E. Franklin, R.O. Welch, J.A. Cross & A.E. Woodal (1974). Resource and environmental profile analysis of nine beverage container alternatives. US Environmental Protection Agency, Washington DC.
- Jelinski, L.W., T.E. Graedel, R.A. Laudise, D.W. McCall & C.K.N. Patel (1992). Industrial Ecology: Concepts and Approaches. Proc. Natl. Acad. Sci. USA, vol. 89, pp 793-797.
- Kandelaars, P.P.A.A.H. (1998). Material-Product Chains: Economic Models and Applications. PhD thesis, Free University Amsterdam.
- Maarel, E. van der & P.L. Dauvellier (1978). Naar een Globaal Ecologisch Model voor de ruimtelijke ontwikkeling van Nederland. Rapport Ministerie van VRO, SDU, Den Haag.
- Opschoor, J.B. (1994). Chain management in environmental policy: analytical and evaluative concepts. In: J.B. Opschoor & R.K. Turner (eds.): Economic Incentives and Environmental Policies, Kluwer Academic Publishers, Dordrecht.
- Organisation for Economic Co-operation and Development (OECD) (1998). Eco-efficiency. OECD publications, Paris.
- Voet, E. van der (1996). Substances from cradle to grave. Development of a methodology for the analysis of substance flows through the economy and environment of a region. PhD thesis, Leiden University.
- Voet, E. van der, R. Huele, R. Flipphi & A. Oosterhof (1996). Biodiversiteit als beleidsconcept. Rapport van de Raad voor het Milieubeheer.
- Weiszäcker, E.U. von, A.B. Lovins & L.H. Lovins (1997). Factor Four - doubling wealth, halving resource use, the new report to the Club of Rome. Earthscan Publications Ltd, London.
- World Commission on Environment and Development (1987). Our Common Future. Oxford University Press, Oxford/New York.

Part II Models for the analysis and management of heavy metals in the Netherlands

Contents:

Part II is dedicated to the models developed in the 'Metals' programme. Section II.1 contains a reflection on modelling issues, addressing the modelling principles employed in economic and environmental/metabolic models and their applicability to the case of heavy metals. In Sections II.2 – II.5 the developed models are described: the general SFA model FLUX in II.2, various integrated chain MPC models in II.3, the detailed SFA model for agricultural soils, DSCB, in II.4, and the general multi-media environmental model Dynabox in II.5. Section II.6 addresses the issue of interpretation: how can the results of the modelling efforts be translated into terms of sustainability and environmental quality? Section II.7 is integrative once again and treats the possibilities of combining metabolic and economic models.

II.1 General modelling issues for economic and substance flow models

Mathijs Bouman

II.1.1 Introduction

Research question 2 of the 'Metals' programme (see Section I.1) concerns the simulation of metal flows and stocks in models. As shown in Part II, a number of models have been developed to address the various aspects of the societal and environmental metabolism, the environmental risks and the management of metals. Although these models are all different, the modelling techniques they use and the modelling issues that play a role are general issues, and decisions about these issues need to be made for every model. This section treats a number of such general modelling issues.

By definition a model is a simplified representation of a part of reality. Since reality itself is a tangled web of cause and effect and interrelations, we can only hope to learn by abstracting from the 'unimportant' and focusing on the 'important'. Described this way, the model-builder's main task is to devise means for simplifying and structuring the overabundant dependencies and interdependencies that can be observed in the real world. The criteria used in this process are determined mainly by two factors. Firstly (and obviously) the purpose of the model is an important determinant for the simplifications that can be made. For a model used to identify the main sources of metal pollution, for example, different simplifications will be made than for a model employed to determine the most efficient policy for tackling pollution problems. Secondly, the choices made in the design of the model are influenced by the model-builder's views. A researcher who believes that heavy metal pollution can only be abated by command and control type policies, for instance, will construct a different model than a researcher who relies on market-based instruments (i.e. taxes and subsidies) to change individual behaviour.

From this perspective, a description of the simplification methods of economic and substance flow models can serve as a useful approach to understanding the essential differences and similarities. Therefore, this section is devoted to the discussion of the techniques used by model-builders in economics and environmental science to simplify observed reality. The main purpose of this overview is to give the reader of this book a feeling for the basic decisions involved in the construction of a model and motivate, in general terms, the simplifying assumptions encountered in the models discussed in the next few chapters.

We distinguish the following six aspects of simplification:
- focus,
- aggregation,
- linearity,
- optimisation,

- time,
- uncertainty.

These topics are treated successively in the following section.

II.1.2 Six modelling issues relevant for economic and substance flow models

Focus

The first and most basic decision that a model-builder makes is to define the boundaries of the system being analysed, by 'zooming in' on a certain part of reality. The choice of boundaries is based on the question(s) the model is supposed to answer. This is done quite literally when the geographical region for the analysis is chosen. Not many models cover the whole world, so in most models assumptions are made about the insignificance of interactions of the region under study with its surroundings. For some models, such as LCA models, a choice is made not to specify boundaries in space. Besides plain geography, the system boundaries may restrict the analysis to a single sector or industry, or a group of sectors or industries. Furthermore, in many cases a decision is made to single out a specific material or substance (e.g. in SFA), a specific product or product function (e.g. in LCA and in many economic partial equilibrium models, including MPC models) or a specific environmental problem (e.g. energy models and climate models). Finally, the model-builder may want to apply the analysis to a certain period in time, or even a certain point in time (see below).

Aggregation

Having set the system boundaries by zooming in on the relevant part of reality, the next step — paradoxically — is to decide how much to 'zoom out'. Since almost all actors, processes and materials in the economy are heterogeneous in one way or another, the model-builder has to decide to what extent they are to be treated as homogeneous. In macro-economic models, for example, economic agents are usually lumped together in such groups as 'households', 'industry', 'government' and 'rest of the world'. Demand functions are summed to form one or more 'aggregate demand functions', while prices are consolidated into price indices. In environmental models substances and materials can be compiled into physical aggregates (e.g. kilograms or Joules) and environmental impacts can be summed up in a single index (e.g. ecological footprint). However done, the aggregation process is unavoidably an arbitrary routine. The choice of aggregation weights has a major impact on the outcome of the model. In the case of the models developed in the 'Metals' programme it has been opted to use indicators to interpret the results of the model calculations. These indicators can be calculated by means of an aggregation routine. They may also be selections. Information from outside the models may be introduced as well, as when using politically determined environmental quality standards.

Linearity

Mathematically, a model consisting of linear relations is easier to solve than one involving non-linear relations. In all types of model, therefore, the evident linearities in real-world relations are gratefully exploited. Examples of such linearities are mass-balance conditions and constant emission factors in MFAs and, in purely economic models, budget restrictions (implying that expenditure equals income). The use of linear relations is not confined to cases where linearity is an apparently valid assumption, however, but can be extended to all the equations of the model. This is what is done in all models based on input-output analysis (IOA) techniques, such as most SFAs and MFAs, economic IOAs and systems of national accounts. In these entirely linear models the emphasis of the analysis is on systematic bookkeeping of observed interactions between the 'nodes' (e.g. agents, sectors) of the model, rather than on explaining these interactions. A linear SFA of cadmium in the Netherlands, for instance, describes the flow of this metal through the economy and into the environment, without claiming to answer questions on why these particular flows take place (Van der Voet, 1994). Put differently, the cadmium SFA shows the cadmium input requirements for the cadmium output of each sector, and assumes that the underlying technology remains unchanged. This means that questions concerning the mechanisms of substitution of inputs and technologies and (dis-)economies of scale cannot be addressed. It seems a high price to pay, but the advantages of strictly linear models are considerable. First of all the mathematics involved in solving the model is relatively simple. This is especially true when the number of relations modelled equals the number of variables, since in that case matrix algebra can be used. Second, the simplicity of linear models allows a large number of relations to be incorporated without rendering practical implementation unfeasible.

Some model-builders opt not to use linear relations. In almost every case this implies that only a small part of reality is covered by the model: there is a trade-off between linearity and focus. The MPC models and the DSCB model for agricultural soils, both of which have a restricted focus, contain non-linear relations for the sake of making model outcomes more 'true to life'.

Optimisation

Most researchers aim for their models to have a single, discrete solution, since for most applications there is no point in calculating a range of possible outcomes. In the aforementioned linear models a single solution is generally guaranteed by modelling as many relations as there are variables. But what if substitution between, say, a metal and a less hazardous alternative is to be modelled? Potentially, when any mix between the two materials is feasible, the possibility of substituting the metal for the alternative may yield a model with an infinite number of solutions. Clearly, a mechanism is needed that determines which mix will eventually be realised. In most cases this mechanism is one that distinguishes the 'best' solution from 'inferior' solutions. In economic models the criteria of maximum profit for firms and maximum utility for households is usually the mechanism that does the trick. From an infinite set of possible input mixes firms choose the one that, given the prices of inputs, yields the highest profits and from a infinite set of possible consumption packages households pick the one that, given product prices, yields the highest utility. Analogously, in environmental models a policy-maker will select the outcome that gives the lowest environmental impact. For instance, in an energy model one may choose the input mix that, given a certain demand for energy, minimises greenhouse gas emissions. This optimisation process to find a single solution is less straightforward than one might think. First, it must be possible to distinguish 'good' solutions from 'bad' solutions. This means that we must be able to order the solutions at least cardinally. For environmental models this is often difficult, since it necessitates some sort of aggregation of different environmental impacts: optimisation therefore requires aggregation. This is not the case for SFA models for heavy metals, however: since only one substance is regarded at a time, there is no need for aggregation. Second, it is necessary for the model to have a single optimum and this criterion often compels the model-builder to make rather artificial assumptions concerning the shape of the model equations – which implies non-linearity. These can often be summarised as an assumption of decreasing marginal returns. For production this means that each additional unit of a given production factor (keeping all other factors constant) adds less to total output than the preceding units. As a result the optimum amount of, for instance, labour employed in production is a finite and unique number. For consumers the assumption of decreasing marginal returns implies that for a given level of consumption of all but one good, each additional unit of that good raises utility less than the preceding units. As for production, this guarantees that the optimal level of consumption for each good is finite and unique. Most of the models from the 'Metals' programme do not employ optimisation but yield a single outcome. The only exception are some of the MPC models described in Section II.3. Theoretically, the SFA models might also be translated into optimisation models, for example providing specifications for management of the metals that keep functions intact while also remaining within statutory concentration limits. Although the goal would here be singular (stay within the limits), there might be a number of different regimes that conform to this.

Time

A very important decision that has to be made in both linear and non-linear models is whether to include time in the analysis. Adding dynamic relations complicates a model considerably, so model-builders generally restrict their models to static relations whenever possible. For some types of model this problem is purely academic. Life cycle assessment, for instance, is concerned with comparison of two alternative products, judging them on the basis of their environmental impact, which is a type of analysis for which a dynamic variant is difficult to imagine – as also holds for space, they abstain from any form of time specification or, as LCA practitioners put it, they integrate over time by specifying all possible future impacts whenever these may occur. For MFA/SFA and economic models the question of time is more relevant. In principle, any model that deals with stock-flow interactions (as is common for both types of models) should stipulate dynamic relations. What would it mean to specify the leakage from a stock of dumped metals into the environment or the depreciation of a capital stock, for instance, without explicit incorporation of time in the analysis? Fortunately, things are not as prohibitive as it would appear from the above. We can construct models that include stock-flow interactions while avoiding the difficulties of dynamic modelling by using the concept of the steady state, defined as the situation in which all flows into stocks are balanced by equal outflows from stocks, so that the stock itself does not change. In economics such steady-state modelling is referred to as comparative static modelling. Clearly, in this equilibrium situation information about the flows and stocks in other years can be neglected. Since it is also the equilibrium into which most well-behaved models will eventually settle, researchers feel comfortable about using the steady-state solution of static models as a means of addressing questions not directly related to the situation of disequilibrium.

Despite the usefulness of the steady-state concept, modelling dynamics cannot be avoided if the research question is inherently dynamic. Examples include questions concerning disequilibria, economic growth, expectations (about future environmental policy, for instance), the build-up of soil concentrations and uncertainty. In the case of metals, moreover, it may take centuries or even millennia to reach this steady state, which is a time horizon quite outside the scope of environmental policy. If some notion is required of what happens between now and eternity a dynamic model is required. All the models developed in the 'Metals' programme also have a dynamic mode, apart from a static 'accounting' and a steady-state mode.

Uncertainty
Generally, in both environmental and economic models the fundamental assumption is made that we are in principle able to model the world with certainty. The results of a model simulation will of course be presented with caution, and prudent researchers will meticulously stipulate the limitations of the analysis, but this is not the same as explicit incorporation of uncertainty in the analysis itself. There are two types of uncertainty that should ideally be included in the model.
The first is uncertainty about what will actually happen in the future. In the context of environmental analysis cases in point include the intrinsic uncertainty about the future impacts of current emissions (e.g. whether or not a runaway greenhouse effect will occur) and uncertainty about the future impact of presently formulated environmental policy. The simple fact that in these cases a certain amount of time elapses between cause and effect gives rise to uncertain outcomes. Dixit and Pindyck (1994, p. 395-421) is one of the few examples of an analysis that takes this type of (environmental) uncertainty into account. The authors show that in the case of 'irreversibilities' the costs and benefits of acting now to prevent possible future damage should be appraised in terms of 'option values'. If a policy-maker decides to implement environmental policy today he foregoes the option of not acting at all (if future research proved to negate the environmental problem). If he decides to postpone policy and wait for a conclusive answer about the environmental problem in question he gives up the option of acting in time. Clearly, comparing the costs and benefits in these terms is a difficult exercise, so it should come as no surprise that in most applications this type of uncertainty is more or less neglected.
The second type of uncertainty is model uncertainty. This term covers all uncertainties related to the description of the world as it is today. This includes uncertainty about the reliability of the collected data, about the correctness of the chosen functional form of the model and about the stability of the described relations. Ideally, a researcher should appraise the probabilities involved in these uncertainties and express the parameters and equations of the empirical model in terms of probability intervals. Again, this would be a difficult and time-consuming operation, which is the reason that in most cases accounting for model uncertainty boils down to calculation of two or more scenarios as a sensitivity analysis. The model applications carried out in the framework of the 'Metals' programme are no exception to this common practice.

II.1.3 Concluding remarks

Any thematic presentation of simplifications like the above might easily give the wrong impression that in practice model-builders freely combine simplification techniques to serve the analysis of the problem at hand. One might erroneously conclude that constructing a model means deciding which simplifications are to be 'switched on' and which 'switched off' — as if the optional simplifications can be combined in any way we fancy. In reality, the simplifying assumptions are often incompatible and construction of a model generally boils down to using a *prescribed set* of assumptions that fits a certain type of analysis. In this manner each modelling discipline has developed its own set of simplifying assumptions to work with. This is of course a practical arrangement serving discussion and progress *within* a discipline, since it avoids needless reiteration of basic assumptions. For the discussion *between* disciplines, however, it poses a problem, since it renders the models from different disciplines incompatible (Bouman et al., submitted). This is best seen by briefly comparing the basic assumptions of substance flow models and economic equilibrium models.

In SFA models the focus of analysis is fairly restricted, since the focus is on a single substance (group) but at the same time rather loose, because this single substance is recorded everywhere in the economy. In contrast, economic equilibrium models tend to include a multitude of inputs, products and actors but focus on only part of the economy. There is an even greater difference in how SFA and economic models use the simplifying method of aggregation. In SFA all variables are expressed in their 'substance contents' (e.g. kg of Cd), which permits aggregation over different products. By this choice of unit, moreover, violation of the law of mass conservation is easily avoided. The outcomes of an economic model is generally expressed in 'value terms' (e.g. dollars, or a vague concept such as 'utility'). Mass conservation laws could in principle be added to these models, but in most cases they are flatly ignored. Evidently, the difference in measurement units is a severe handicap to integration of the two types of models.

The same is true of the methods used to guarantee a single solution to the model. Almost without exception, SFA models contain only linear relations. As explained above, this allows the model-builder to include many relations (as is necessary if one wants to model a substance from cradle to grave) at the expense of foregoing the possibility of examining substitution mechanisms. In economic equilibrium models a single solution is obtained by using optimisation techniques. This permits explicit discussion of substitution mechanisms. The non-linearities required to warrant a single solution complicate the analysis to such a degree that the use of this type of model is confined to either small parts of the economy, or highly aggregated economies. Since the choice for linear or non-linear modelling greatly affects both the structure and the scope of the models, integration of SFA and economic equilibrium models is very difficult. In Section II.7 we treat this issue in more detail.

References

- Bouman, M., R. Heijungs, E. van der Voet, J. van den Bergh & G. Huppes (1998). Material Flows and Economic Models. Ecological Economics, in press.
- Dixit, Avinash K. and Robert S. Pindyck (1994). Investment under Uncertainty. Princeton University Press, Princeton NJ.
- Heijungs, R. (1997). Economic Drama and the Environmental Stage - Formal Derivation of Algorithmic Tools for Environmental Analysis and Decision-Support from a Unified Epistemological Principle. PhD thesis, Leiden University.

II.2 FLUX, a tool for substance flow analysis

Xander Olsthoorn & Jos Boelens

II.2.1 Introduction

This chapter presents a concise description of a software tool (**FLUX**) that has been developed for the analysis of patterns of materials use (physical flows and physical stocks) in an economy and of associated patterns of environmental pollution. This type of analysis is known as substance flow analysis (SFA) and as materials flow analysis (MFA) (see e.g. Vellinga et al., 1998; Schmidt and Schorb, 1995). SFA may be compared with economic analysis (e.g. input-output modelling) that is based on national accounts that describe economies. While national accounts map the flow of goods and services measured by their prices, SFA measures these flows by their contents of a given substance. A substance flow account - a chemical cross-section of the flows and stocks in an economy and in the environment (Olsthoorn, 1991) - is the core of SFA and the starting point of modelling. An important characteristic of SFA is that it aims to be chemically comprehensive, taking account both intended and unintended uses (flows) of substances (e.g. cadmium in foodstuffs).

A software tool for performing SFA is useful for various reasons. Firstly, such a tool can facilitate the groundwork for SFA, i.e. create its empirical basis. Experience (e.g. Thomas et al., 1985; Wernick and Ausubel, 1995; Van der Voet, 1996) shows that finding and processing suitable statistical information for establishing values for flows and stocks is a tedious job, since sources of appropriate information vary widely in nature and data are often mutually incompatible or inconsistent. A properly designed database can be helpful in managing this task. In conjunction with the database function, dedicated software can facilitate analysis of the data and modelling. A third reason is that software can facilitate the linking of SFA information to other types of models, e.g. economic models.

FLUX supports different tasks that have to be performed in the context of this analysis, ranging from database functions to supporting dynamic modelling. FLUX has been applied to a series of heavy metals (copper, lead, cadmium and zinc), aluminium and (organic) carbon. As far as we know (Boelens and Olsthoorn, 1998; Schmidt and Schorb, 1995) FLUX is the first tool that combines all these functionalities.

In the following we first describe the scope of FLUX. Section II.2.3 considers the design principles and the design of FLUX. Sections II.2.4 and II.2.5, respectively, summarise results and discuss links to other models.

II.2.2 Goal and scope of FLUX

The basic concept of SFA on which the design of FLUX is based is the idea of a chemical cross-section of the physical economy-environment system in some geographical area (Van der Voet, 1996; Olsthoorn, 1991). This cross-section is viewed as a network that comprises (*i*) nodes, (*ii*) flows of a selected chemical substance between nodes and (*iii*) stocks of that substance, held by the nodes. For each node, the law of mass conservation must apply. This concept distinguishes nodes in two domains: the economic system and the ecological system. An economic node refers to an economic entity as distinguished in economics (e.g. a firm, an economic sector). Ecological nodes (or environmental nodes) refer to parts of the ecological system (e.g. atmosphere, soil, water). The substances that 'flow' into a node are subjected to a **material transformation** in that node. In economic nodes transformation refers to chemical or physical processes that are managed by humans, while in ecological nodes transformation refers to the dispersion of substances, a process governed by the laws of nature. This difference in the kind of transformation constitutes the main distinction between the two types of node.

Against the background of this model of the physical economy and its adjacent environment, SFA addresses different kinds of environmental question (see Table II.2.1). The FLUX software supports all the associated analysis. So far it has been used for SFAs of the four heavy metals lead, cadmium, zinc and copper in the Netherlands.

Table II.2.1 Questions from different perspectives for substance flow analysis: the context of FLUX.

Perspective	Questions
Environmental analysis	How do substances enter the environment? Which stocks may pose environmental risks? How do problems develop over time?
Economics	Which goods contain the harmful substances? What is the economic values of each of these goods? In what markets are they sold and traded? And which economic sectors are involved?
Technology	What technologies are used in the processing of relevant goods, and what are the technological alternatives? What is the effectiveness and cost of technologies to control the environmental problems?
Policy-making	Which actors (sectors) are important? What parts of the system (economic sectors, technologies) provide effective leverage for policy tools? What constitute useful indicators for policy-making and monitoring?

FLUX was built to help provide answers to such - but not all - questions within the context of the research reported in this volume. FLUX is in fact designed to support the following tasks:

- compiling statistical data on the uses of materials and their chemical composition, and keeping a database for retrieval of this information;
- describing networks of nodes and flows of different substances in a consistent manner and characterising them by means of indicators;
- performing database functions that support analysis;
- constructing solvable and meaningful models that represent substance flows and stocks and their development over time;

- simulating the behaviour of the substance flow network over time for performing policy scenario analysis.

The sectoral scope of FLUX is not limited to flows in 'the economy'. FLUX may also be used to describe stocks and flows in the environment, in a similar way to first-order fate modelling for assessing the risks of hazardous substances (Van de Meent et al., 1995). In the present research FLUX was not used in this way.

An important characteristic of FLUX is that the system permits dynamic modelling, that is it can calculate future substance flow accounts. Input data for such calculations are: scenarios for the development of flows that are assumed to be drivers of the system (e.g. 'final demand' flows) and also scenarios for technological change (technological coefficients) and the time horizon.

FLUX is designed for use in conjunction with other type of models, e.g. economic models and environmental models. The nature of the link between FLUX and these other models is primarily to facilitate data exchange between these models.

FLUX has been built from the Windows™ version of the FoxPro™ relational database management system.

II.2.3 Modelling principles and data requirements

FLUX objects

We have adopted an object-oriented approach in developing a software system for substance flow analysis, focusing on the following five objects: goods, substances, nodes, flows and stocks. Objects are defined by their attributes. **Goods** are the actual matter that 'flows' between nodes. The principal attribute of a good is its composition. Composition can be expressed in two ways: either a good may have a defined number of subparts (a car having five tyres) or the chemical composition of the good may be specified (a car contains copper, in kg Cu per car; rubber contains zinc, in kg Zn per kg rubber). **Substances** are chemical constituents of the goods (e.g. zinc in cars). FLUX contains a database that permits entry and retrieval of data on the chemical composition of goods (e.g. the lead content of fertiliser, the cadmium content of zinc ore). **Target substance** are the selected substances (e.g. elements such as lead, cadmium or carbon) for which the modelling is being carried out.

'Sector' may refer to either economic or environmental sectors, with its meaning obviously differing in these two domains. Similarly, the meaning of '**process**' differs in the economic and the environmental domain. For nodes in the former domain, 'process' refers to the technology according to which the materials are transformed by an (economic) sector. In the environmental domain it indicates, for instance, the type of dispersion process (e.g. dry and wet deposition). **Location** is the geographical attribute.

The actual sectors, processes, locations, goods and substances must be defined by the user of FLUX. These objects constitute the **building blocks** of FLUX. The very first step when using FLUX is to define these building blocks. Each of these objects can be ordered on the basis of a taxonomy, also to be defined by the user of FLUX. For

instance, one may assign a sector 'fertiliser manufacture' to a category of 'chemical industry', which, in turn, can be assigned to a category 'manufacturing industry'. This functionality of FLUX permits filtering of the database.
Proper design of these building blocks goes a long way to determine the potential for using FLUX results in other types of models. For instance, using the classification system employed in national economic accounts to categorise economic sectors ultimately facilitates the use of FLUX data in economic models.

Nodes have three main attributes: they are termed sector, process and location. Together, they uniquely define a node in FLUX. Table II.2.2 gives some examples. The attributes that uniquely define a **flow** are: the source node, the destination node, the name of a good/material, its flow rate (e.g. tonne per year) and the year to which the information pertains. In addition, FLUX records the uncertainty of the flow. **Stocks** refer to the amounts of a specified substance in material assets at a particular point in time. Attributes of stocks are the node, the name of the good/material, its magnitude, year and uncertainty.
Nodes, flows and stocks can be characterised. These characteristics are used for environmental classification of flows (e.g. extraction, production, waste processing). These characteristics and their use in establishing indicators are discussed elsewhere in this book.
Since FLUX is designed to function as a database, there is ample scope for annotating entered information with comments and references.

FLUX procedure
FLUX supports a number of tasks that are part of SFA. Figure II.2.1 shows these steps, which characterise analysis of the flows and stocks of a substance.

Figure II.2.1 Steps in performing SFA that are supported by FLUX.

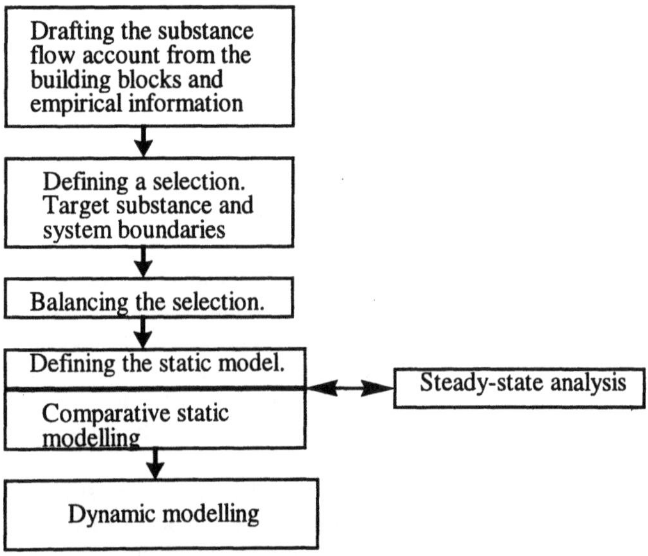

Drafting the substance flow account

The first step in performing SFA with FLUX is construction of a substance flow account from whatever information is relevant for estimating flows and stocks. This step is actually threefold: (*i*) define the nodes, (*ii*) define the materials that flow and their chemical composition and (*iii*) enter the data on flows (and stocks) in FLUX. The nodes are defined from the lists of sectors, processes and locations (the building blocks of FLUX). As an example, Table II.2.2 lists some of the constituent nodes of the database used for studies elsewhere reported (Guinée et al., 1998). Nodes printed in italics are assumed to comprise stocks. Typically, a heavy metal substance flow account (Guinée et al., 1998) has over 50 nodes.

Table II.2.2 Examples of nodes distinguished in FLUX applications. Nodes that contain stocks are in italics.

Node ID	Sector (actor)	Process	Location
5	Sewage works firms	Aerobic sewage water treatment	The Netherlands
9	Non-ferrous primary metals industry	Hydro metallurgical zinc refining	North Brabant Province, the Netherlands
40	Livestock farming	Feeding animals	The Netherlands
86	Inland shipping	Wear & corrosion	The Netherlands
170	Road infrastructure (Public authorities)	Use of bulk building materials that contain copper (e.g. cement)	The Netherlands
57	Agricultural soils	Natural processes	The Netherlands

FLUX allows comments to be added to the numerical information. In addition, when flow data are entered FLUX asks for an indication of the uncertainty of the information (to be used in the balancing step; see below).

Defining a selection

The second step is to select a target substance and cut the corresponding chemical cross-section of the data and define the boundaries (which nodes to include or exclude). This is called the selection step. The result of the step is an account of the flows and stocks of the selected target substance. Actually, it is in this step that FLUX creates a flow account at the substance level, by recalculating flows and taking account of the chemical composition of the materials. For instance, the flow of iron ore into the node 'Primary metals industry' is recalculated as a flow of zinc to this node, using the data on the zinc content of iron ore. In other words, FLUX selects a substance flow account from the materials flow account.

Balancing the selection

It is difficult to find appropriate empirical information on substance flows and stocks. Hardly any data has the required format, the data is often ill-defined and there are major uncertainties. Given the uncertainty of many data, it is unlikely that the initial information will allow an account to be drawn up that is mass-balanced at each node. Many nodes are thus likely to be unbalanced owing to data uncertainties. Balancing the nodes simultaneously may be time-consuming, because of the mutual dependencies of the balances, which must be duly accounted for. We therefore developed a procedure for balancing the entire network, reducing mass imbalances at all nodes simultaneously using 'uncertainty classes' attributed to the flow data, each class representing a range of uncertainty values. The balancing problem is identified as a so-called discrete optimisation problem. FLUX solves this problem by iteratively minimising:

$$P_f = f\sum_p \left(I_p - O_p\right)^2 + \sum_i g_i \left(w_{in} - w_{io}\right)^2 \qquad (0.1)$$

Where

P_f = overall measure for the 'balance of the network'

f = an adjustable factor that controls the balancing procedure

p = index for nodes

I = Input

O = output

g_i = a weighting factor that is derived from the uncertainty in flow i. Under large uncertainty g_i (adjustable) will relatively small.

$\left(w_{in} - w_{io}\right)$ = the adaptation of flow i. w_{io} is the old value, w_{in} is the new value

After successful balancing the condition of mass conservation is better met, at the expense of modification of the initial data on the flows of the target substances. However, these modification are within the ranges of uncertainties of the flows.

Balancing is a purely mechanical procedure that helps improve a substance flow account with respect to obeying mass conservation requirements. Balancing can never be complete in the sense of each mass balance being precisely struck, even if an infinite number of iterations were to be performed. The FLUX user controls the length of the procedure and the final result by defining the number of iterations and the control variable f.

Static modelling

The fourth step is specification of the relations between the flow and stocks of the substances. The resulting model is a mathematical representation of the structure of a substance 'stocks and flow' account for the reference year, taking account of the mutual dependencies of the flows. The principal assumption of the model is that flows depend linearly on other flows or on stocks. Flows may be proportional to:

- the total output (throughput) of its destination node (by definition a flow has a source node (origin) and a destination node);
- the total input (throughput) of its source node;
- the stock that constitutes its source;
- a balance item in the mass balance of a node (for example, use of a virgin feedstock depends on the total use of feedstock minus the available amount of secondary feedstock).

The first dependency in fact refers to the fixed-technology assumption of Leontief-type input-output modelling and applies to flows into 'economic' nodes. The second dependency reflects the fact that flows into environmental nodes depend on total source inputs (e.g. an emission is assumed to be proportional to total input). The third assumptions states that some flows depend linearly on a stock; for instance, the flow of car scrap is assumed to be proportional to the total stock of cars. Another example is that the rate of leaching of a substance from a landfill is assumed to be proportional to the stock of that substance in that landfill. Finally, in some cases flows must be considered dependent on neither destination output nor source input, but rather as the outcome of a balance. For example, the use of virgin pulp in the paper industry may be assumed to be the result of balancing the total demand for pulp and the supply of pulp from recycled paper. In FLUX the latter flow could be considered to be the result of a balance over a destination node.

Figure II.2.2 shows an example of a network of flows and stocks for which we want to build a model, with x_i representing flows, N_i stocks and x_1, x_8 and x_9 flows to/from beyond the system's borders.

Figure II.2.2. An example of substance flow system; the shaded circle represents a node containing a stock.

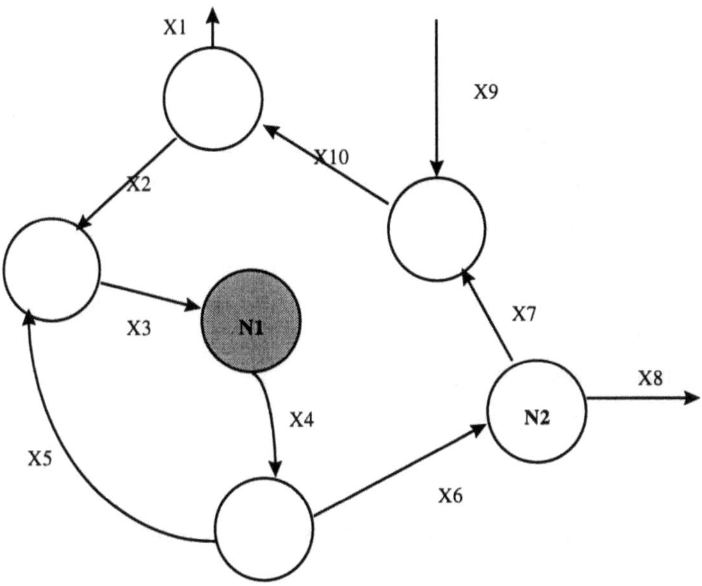

We consider X1, X3 and N1 to be the independent variables and our aim is to find all other x for some set of values for these variables.

We can write the following (linear) mass balance equations:

X1 = a	Independent (X1 would be export, for example)
X3 = b	Independent
N1= c	Independent

$X2 - X10 + X1 = 0$	Balance equation
$X4 - \beta * N1 = 0$	A flow driven by the stock (e.g. the generation of scrap)
$X5 - \alpha * X4 = 0$	X5 dependent on source throughput (α is recycling rate)
$X6 - (1-\alpha) * X4 = 0$	X6 dependent on source throughput
$X8 - \gamma * X6 = 0$	X8 dependent on source throughput (γ would be emission coefficient)
$X7 - (1-\gamma) * X6 = 0$	X7 dependent on source throughput
$X9 - \delta * X10 = 0$	δ is an input coefficient (dependent on destination throughput)
$X10 - X7 - X9 = 0$	Balance equation.

These equations are specified using a FLUX function that allows the user to indicate, for each flow, what type of dependency pertains. FLUX calculates the coefficients from the - balanced - values of the flows and stocks. For instance, if $\alpha = 0.33$, $\beta = 0.0001$, $\gamma = 0.0025$ and $\delta = 0.2$, these equations can be written in matrix form as follows:

1	0	0	0	0	0	0	0	0	0	0		a		N1
0	1	0	0	0	0	0	0	0	0	0		b		X1
0	0	1	0	0	0	0	0	0	0	1		c		X3
0	1	0	1	0	0	0	0	0	0	-1		X2		0
-0.0001	0	0	0	1	0	0	0	0	0	0		X4		0
0	0	0	0	-0.3300	1	0	0	0	0	0	*	X5	=	0
0	0	0	0	-0.6700	0	1	0	0	0	0		X6		0
0	0	0	0	0	0	-0.9975	1	0	0	0		X7		0
0	0	0	0	0	0	-0.0025	0	1	0	0		X8		0
0	0	0	0	0	0	0	0	0	1	-0.2000		X9		0
0	0	0	0	0	0	0	-1	0	-1	1		X10		0

In summary matrix notation the model is:

$\mathbf{A}.\boldsymbol{x} = \boldsymbol{y}$

where:

- $\boldsymbol{y}$ is the vector containing the selected independent variables (in the example: x_1, x_3 and N_1),
- $\boldsymbol{x}$ is the vector of dependent variables (flows and stocks), and
- **A** is the matrix of coefficients.

FLUX asks the user to indicate the dependencies of the various flows and indicate which flows are to be considered as independent variables. FLUX then creates the model (A and A^{-1}) after checking the consistencies of the indicated dependencies in order to form the basis for a solvable model. Next, after the user has entered a series of values for the independent variables - that is, having defined a vector y - the flows and stocks (both in x) are calculated according to $\mathbf{A}^{-1}.\boldsymbol{y} = \boldsymbol{x}$.

The example's A^{-1} - as calculated by FLUX - is shown below.

1	0	0	0	0	0	0	0	0	0	0
0	1	0	0	0	0	0	0	0	0	0
-0.000084	0	1	0	-0.8354	0	-1.2469	-1.2500	0	-1.2500	-1.2500
0.000084	-1	0	1	0.8354	0	1.2469	1.2500	0	1.2500	1.2500
0.000100	0	0	0	1	0	0	0	0	0	0
0.000033	0	0	0	0.3300	1	0	0	0	0	0
0.000067	0	0	0	0.6700	0	1	0	0	0	0
0.000067	0	0	0	0.6683	0	0.9975	1	0	0	0
-0.000001	0	0	0	-0.0067	0	-0.0100	-0.0125	1	-0.0125	-0.0125
0.000017	0	0	0	0.1671	0	0.2494	0.2500	0	1.2500	0
0.000084	0	0	0	0.8354	0	1.2469	1.2500	0	1.2500	1.2500

FLUX allows A to be changed before calculations are performed, thus permitting examination of the effects of technological change as represented by changes in coefficients.

Dynamic modelling

FLUX also allows dynamic modelling, that is modelling in which time is explicitly accounted for and the development of flows and stocks over time are calculated. Inputs to the model are (*i*) scenarios (time paths) for the development of selected independent variables ($y(t)$ comprising, for instance, a time path for final agricultural demand for phosphate fertilisers) and (*ii*) scenarios for the development of matrix A, the coefficients. The calculations are technically straightforward: for each year of the chosen scenario period FLUX calculates

$\{A(t)\}^{-1} . y(t) = x(t+1)$.

The variable $y(t)$ includes all stocks. Next, the stocks of $y(t+1)$ are calculated from the stocks at time t and the mass balance deficit/surplus of the node at t. The calculation is then repeated until the final year of the scenario period.

Steady-state modelling

A special form of dynamic modelling is steady-state analysis. This type of analysis asks: What happens to stocks when both $A(t)$ and y(t) are kept constant over an infinite time? Obviously some of the stocks will increase and others will decrease, and eventually a steady state is reached when, for each node, inputs equal outputs. For instance, consider a single node with a (constant) input flow I, a stock $N(t)$, and an output flow $O(t)$. $O(t)$ is proportional to $N(t)$:

$$O(t) = k * N(t)).$$

$O(t)$ does not equal I, thus $N(t)$ will change over time according to:

$$\frac{dN}{dt} = I - kN(t)$$

The solution of this equation is:

$$N(t) = e^{-kt}\left(N(0) - \frac{I}{k}\right) + \frac{I}{k}$$

where $N(0)$ is the initial stock. Whether initially the stock will increase or decrease depends on the values of k, I and $N(0)$. For $t = \infty$ the latter equation results in:

$$I = kN(\infty) = O(\infty),$$

which situation is called the steady state. Calculation of this state for substance flows systems is called steady state analysis (Van der Voet et al., 1995). Note that it is not necessary to solve the differential equation to find the steady state, since $N(\infty) = \frac{I}{k}$. Next to the input we only require a value for k, which can be interpreted as the inverse residence time (lifetime) of a substance in a node.

To find the steady state of complex substance flow systems, FLUX rewrites $\mathbf{A}x = y$ as a set of equations by replacing all equations that state an output flow being proportional to a stock of a node by equations that state inputs equal outputs of those nodes. Given a set of independent variables, FLUX calculates a vector y that represents the value of the flows that would be attained after infinite time, which is a state of equilibrium, since values of stocks and flows then no longer alter. In the case of the example of Figure II.2.2, performing a steady-state analysis means saying that N1 is a dependent variable and replacing the third of the listed equations (N1 = c) by the equation $x3 = x4$.
Steady-state analysis is explained here as a form of dynamic modelling. However, it may also be considered a special type of static modelling (it is in Figure II.2.1), since time need not be taken into account in actual calculations (as is the case in FLUX). No time paths are calculated, but rather the situation that would evolve after infinite time.

What is the significance of a steady state? In a way, a steady state is an indicator for an economy and its environment, conveying an impression of the direction into which the system of flows and stocks is evolving under a given set of conditions (constituted by assumptions on independent variables (vector y) and on coefficients (A)). Such sets of conditions have been called substance flow *regimes* (Voet et al., 1995). The steady state can be seen as the fingerprint of an economy under such a flow *regime*.

II.2.4 Results

FLUX has been used for research of which the substantial results are presented elsewhere in this volume. The present chapter has concentrated on the methodological aspects of SFA, in connection with practical issues (e.g. database building). With respect to these topics we have some conclusions.

There is a need for software suitable for information management, analysis and modelling. FLUX has been developed to perform these tasks in a single system. FLUX supports modelling of substance flow systems, both comparative-static and dynamic. The specification of the models is supported by a network balancing procedure that facilitates the construction of substance flow accounts that better meet the condition of mass balance.

More or less simultaneously with the construction of FLUX, data on flows and stocks of the heavy metals lead, cadmium, zinc and copper was entered into the evolving FLUX. To a large extent this information was based on earlier studies (Annema et al., 1991). This study as well as earlier studies (e.g. Gilbert et al., 1992; Thomas et al., 1985; Voet, 1996) have focused on individual substances. In FLUX, however, information on individual substances is mutually linked, through the chemical composition of the materials containing these substances (e.g. heavy metals in imported animal fodder).
In defining and characterising the nodes and flows that are relevant to the accounts of these four heavy metals care was taken to describe the nodes and flow in such a way that the information is meaningful to analysts with different disciplinary backgrounds. This facilitates the use of FLUX information and results in other contexts, for instance in economic studies.

Development of FLUX is not yet finished, in particular with respect to the user-friendliness of FLUX, and there is certainly room for improvement (Boelens et al., 1998). However, SFA will continue to be a complex activity, since it aims to be chemically comprehensive in its description of flows in an economy and the adjacent environment. The quantity and variety in the nature of the empirical data that must be taken into account and interpreted will continue to require expertise that cannot readily be incorporated into software tools.

II.2.5 Links to other models

A model built with FLUX aims to comprehensively describe the flows and stocks of some substance in a given area, covering both intentional and unintentional flows and stocks of that substance. Ultimately, these models should provide information about future environmental risks and about approaches to addressing these risks. However, the effort to be geographically and economically comprehensive is at the expense of the detail and scope of the environmental information. For instance, FLUX does not inform on substance-oriented environmental policies that are based on the use of economic instruments. In order to facilitate use of FLUX information in economic models, care was taken to produce data that matches the data formats (in particular, sector classification) used in economic models. FLUX results could consequently be used in a study of the economic effects of materials policies carried out using an applied general equilibrium model (Kandelaars and Dellink, 1997). The environmental risks associated with current patterns of both intended and unintended use of heavy metals have been analysed using the Dynabox model (see Chapter II.4). Finally, FLUX provides information that can be condensed into indicators that environmentally characterise usage patterns of substances (e.g. heavy metals; see Chapter II.6).

References

- Annema, J.A., E.M. Paardekooper, H. Booij, L.F.C.M. van Oers, E. van der Voet and P.A.A. Mulder, (1995), Stofstroomanalyse van zes zware metalen - Gevolgen van autonome ontwikkelingen en maatregelen, RIVM, rapport no. 601014010, Bilthoven
- Boelens, J. and X. Olsthoorn (1998), Substance Flow Analysis with Flux, IVM-VU, Working paper, W98-15, Amsterdam
- Gilbert, A.J. and J.F. Feenstra, (1992), An indicator of sustainable development: diffusion of cadmium, IES, Vrije Universiteit, R92/06, Amsterdam
- Guinée, J.B., J.C.J.M. van den Bergh, J. Boelens, P.J. Fraanje, G. Huppes, P.P.A.A.H. Kandelaars, Th.M. Lexmond, S.W. Moolenaar, A.A. Olsthoorn, H.A., Udo de Haes, E. Verkuijlen and E. van der Voet, Evaluation of risks of metal flows and accumulation in economy and environment. Accepted by Ecological Economics.
- Kandelaars, P.P.A.A.H. and R.B. Dellink (1997) Economic effects of material policies: combining an applied equilibrium model with material flows. Tinbergen Institute T 97-118/3, Amsterdam.
- Kneese, A.V., R.U. Ayres and R.C. d'Arge (1970) Economics and the Environment: a Material Balance Approach. Baltimore: Johns Hopkins Press
- Olsthoorn, X. (1991), Sources of persistent micropollutants: analysis with dynamic materials balances, In Opschoor H. and D. Pearce (Eds) Persistent Pollutants: economics and policy, Kluwer, Dordrecht p. 9-19
- Meent, D.van de, J.H.M. de Bruijn, F.A.A.M. de Leeuw, A.C.M. de Nijs, D.T. Jager and T.G. Vermeire (1995), Exposure Modelling, in C.J. van Leeuwen and J.L.M. Hermens eds. Risk Assessment of Chemicals: An Introduction, Kluwer, Dordrecht, p.103-145
- Perrings, C., (1987), Economy and Environment. A theoretical essay on the interdependence of economic and environmental systems, Cambridge University Press
- Schmidt, M. and A. Schorb ed. (1995) Stoffstromanalyse in Õkobilanzen und Õko-Audits, Springer, Berlin
- Vellinga, P., F. Berkhout and J. Gupta Eds. (1998) Managing a Material World. Perspectives in Industrial Ecology, Kluwer Dordrecht
- Thomas R. and Olsthoorn A.A., (1985), Environmental pollution by lead 1981-2000 (in Dutch), IVM-VU, RIM-16
- Voet, E. van der, R. Heijungs, P. Mulder, R. Huele and L. van Oers (1995), Studying substance flows through the economy and environment of a region - Part II. Environmental Science and Pollution Research, Vol. 2, p. 137-144
- Voet, E. Van der (1996), Substances from cradle to grave. Development of a methodology for the analysis of substance flows through the economy and the environment of a region, PhD thesis, Leiden University, Leiden

II.3 Economic analysis of Material-Product Chains

Patricia Kandelaars & Jeroen van den Bergh

II.3.1 Introduction

Physical aspects of environmental problems are studied by natural and environmental scientists. However, their studies usually do not consider the economic and behavioural mechanisms underlying physical processes and material flows in the economy. In policy design, physical and technological possibilities need to be combined with economic aspects and behaviour. Traditionally, environmental economics has mainly focused on a partial analysis of environmental problems, resulting in a neglect for the interdependence of environmental problems caused by different economic stages. Often environmental economics focuses on external effects, without considering the material or physical dimension of problems. Economic processes are linked to - and even regarded as embedded in - physical processes and therefore a change in an economic process affects the physical process and vice versa. To include this physical dimension, material flow models may be combined with economic models. This allows the study of policy packages in which physical and economic aspects are considered simultaneously. This results in analyses that are economically consistent and physically feasible.

Material-product (M-P) chain analysis tries to integrate physical and economic aspects of material and product flows. It presents an approach that takes the interactions between depletion and pollution into account and regards the economy as being composed of various stages or activities between extraction and emissions. This involves linking the economic and physical aspects of the use of materials. More in particular, M-P chain analysis aims to contribute to integrated model-based analyses of resource and pollution problems for policy making. The approach here is based on the concept "material-product (M-P) chain". An M-P chain can be defined as a set of linked flows of materials and products so as to fulfil a certain service (Opschoor, 1994). An analysis of an M-P chain can be defined broadly as an analysis of the structure of connected material and product flows. An economic analysis of an M-P chain focuses on economic aspects like allocation of products and resources, substitution, recycling and behaviour. Such economic modelling of M-P chains requires combining the elements of physical flow and economic models (Kandelaars, forthcoming).

M-P chain analysis can provide insight into the flows of various materials and products, their interactions, and the impact of implementation of chain policies. This makes it possible to use models of M-P chains for analysis and sometimes even for predicting the effects of management and public policies, technological development and changes in demand for products or materials.

II.3.2 Goal and scope of the models

Research on material flows in economic systems has hitherto mainly focused on describing physical flows in a certain period and region, or related to a particular product. Little attention has been devoted to the economic aspects of physical flows. Economic M-P chain models attempt to fill this gap between environmental science, on the one hand, and economics, on the other. The goal is to examine the physical and economic mechanisms related to flows of materials and products, and the possible policies and strategies to ameliorate the environmental problems associated with these flows.

With the concept of an M-P chain various analyses can be performed. A broad definition of "M-P chain analysis" encompasses both economic and environmental analyses of an economic structure of connected material and product flows. Life-cycle assessment (LCA) can be seen as a case of an environmental M-P chain analysis, because it examines the environmental impact of a product and its material flows. However, an M-P chain is not at the basis of a material flow analysis (MFA) as products are not explicitly described. In this chapter the term a narrow definition of "M-P chain analysis" is used. This narrow definition refers to the study of allocative and economic processes of an M-P chains. Such M-P chain analysis allows the study of material allocative processes, material policies, market equilibrium, market processes, substitution of materials in production functions, and behaviour of agents affecting material use.

Figure II.3.1 shows an M-P chain. Consumers choose between the substitute products A and B. These products may be re-used after they are disposed off. Otherwise they are transformed into the materials of which they consist. This material waste may be recycled or dumped. The products are made of materials 1 and 2. These materials may be either new or recycled materials.

Figure II.3.1 An example of a Materials-Product Chain.

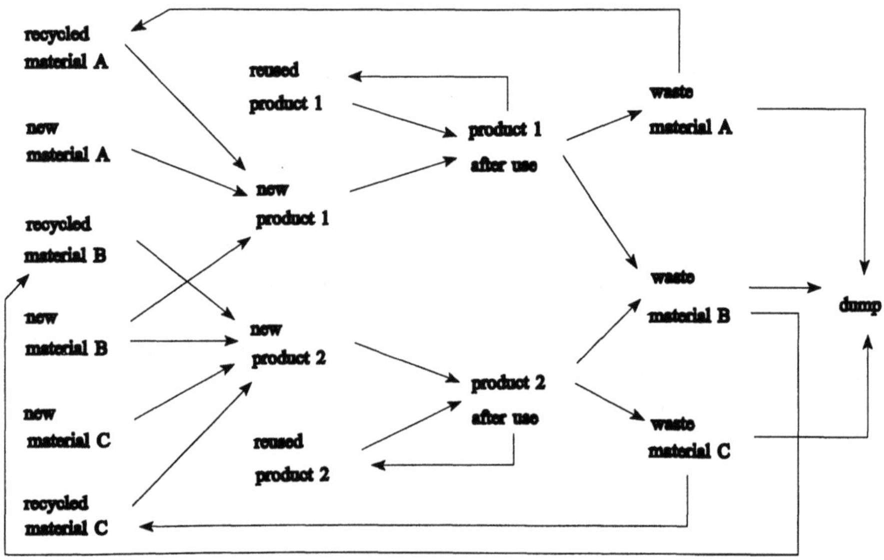

It may be noted that a broad definition of M-P chain analysis includes LCA as a special case. In a more narrow definition as adopted here LCA is not contained, because LCA does not describe the behaviour of individuals, market processes, or optimised allocation by economic agents of resources and products. In other words, it does not deal with economic processes such as studied by economists. Economic M-P chain analysis integrates economic with material flow models, and it takes into account material balance conditions. It therefore aims to link the description of economic processes as studied by economists and that of physical processes in the economy such as studied by environmental scientists. Therefore, in the narrow definition of M-P chain analysis LCA may be used as a basis, but LCA needs to be elaborated with economic processes.

Chain management is based on a mix of instruments that need to be attuned, given the environmental or external effects of each activity in a chain. In M-P chain analysis the linkages of particular activities between resource extraction and waste treatment are considered, allowing also indirect effects of policies to be considered. For instance, a reduction in the use of one material to reach a certain level in terms of environmental indicator X may require less use of a particular product, but then due to a resulting increase of the use of another product, which provides the same service, the use of another material may increase and environmental indicator Y may be negatively affected. These are difficult trade-offs, but they can only be made explicit after the different physical and environmental dimensions are linked to each other via economic mechanisms. In theoretical or analytical economic models of M-P chains these linkages may be included by, for example, extending economic models. For empirical models this is more complicated, because information is required on the behaviour and choices of economic agents with regard to products, materials, recycling and waste treatment. Furthermore, models are required for recycling activities and waste treatment.

II.3.3 Modelling principles and required data

A broad array of model types is consistent with economic M-P chain analysis. The approach essentially combines or integrates - formally or heuristically - an economic model with a physical flow model. Table II.3.1 shows a typology of economic and environmental economic models with some core features summarised. These model types will be referred to later on in section III.3, where some examples of actual integration of these models with physical flow models has been attempted, in various ways.

Materials balance (MB) conditions need to be included for every economic activity to ensure that the economic model does not generate policy options that are physically incorrect. It is also important to keep track of the materials that are contained in a product with a view to the ultimate waste treatment of products that are disposed of. When a production process (or function) has multiple inputs or multiple outputs, it may be a problem to assess the amount of (different types of) material contained in the product. This consideration is also relevant for the transformation of products into materials, because the amount of materials that may be recycled and the quality of the materials may depend on the amount of capital or labour that is used in the transformation process.

The model may be static or dynamic. The choice of the time horizon in dynamic models depends on the goal of the analysis. A dynamic analysis of M-P chains makes it possible to study delayed effects, accumulation of materials or products, technological changes and other development paths. Especially when durable products are considered, a dynamic analysis is more appropriate. The effect of a policy may be delayed: for example, a policy imposed on the material used in production in order to change the amount of waste generated by disposed products, may have a delayed effect when the product has a lifetime of several years (i.e. a durable product).

Table II.3.1 General characteristics of economic model types.

	Economic models of natural resources	Pollution models	Input-output models	Macro-economic models	Models of technological change and economic evolution
Focus	Optimal allocation of natural resources over time	Optimal regulation of polluting activities	Sectoral structure; interaction, indirect and multiplier effect of changes in demand and production technology	Forecasting or scenario analysis of macro-impacts of environmental policies	Irreversible change, gradual and discrete jumps, co-evolution of environment and economy
Theoretical framework	Neo-classical; micro-economic	Neo-classical; micro-economic; externalities	Leontief production function for primary and intermediate inputs; meso-level	(Disequilibrium) relationships between macroeconomic variables	Non-mechanistic, disequilibrium, non-deterministic, learning
Temporal features	Dynamic	Static / dynamic	Mainly static	Dynamic	Dynamic
Model type	Optimisation	Optimisation; equilibrium	Descriptive (optimisation)	Descriptive; forecasting	Descriptive

An economic analysis of M-P chains allows the study of, for instance, optimisation, market equilibrium, policy analysis and scenarios for future development. M-P chain analysis may render insight into the reduction of the environmental burden caused by the demand for a service. An economic analysis of M-P chains may include one or more of the following aspects:

- *Economic processes*, such as prices and costs of materials and products, market equilibrium, allocation, production functions.
- *Behaviour* of consumers and producers when this influences the use of materials and products.

- *Recycling* of materials and reuse of products is, if relevant, fully included. Recycling depends on economic and technical processes
- *Substitution* between different materials between materials and other inputs.
- *Dynamic aspects* such as technological change, accumulation of materials, durable products.
- *Policies and strategies*, such as focusing on recycling, reuse, dematerialization, substitution, waste treatment.

Ideally, the focus of M-P chain analysis is on all environmental aspects. This means that all feasible alternative materials, technologies and products may be taken into account. However, this goes beyond what is practically possible in analytical and data terms. Hence, instead of using "complete" M-P chains, for the purpose of this analysis these are usually "truncated", i.e. an M-P chain is reduced based on economic, physical and environmental aspects, and on data availability (Opschoor, 1994). M-P chain analysis is, like LCA and MFA, limited by data availability and unpredictable future flows. Moreover, certain criteria are needed to truncate all the related material and product flows, and assumptions need to be made on the uncertainty of prices and the impact of policies on consumer and producer behaviour.

By way of illustration a general formulation is offered here of an optimisation oriented model of M-P chain where a service is considered that can be provided by two alternative production technologies. An empirical illustration of this type of model is documented in section III.3.2. The graphical representation is shown in Figure II.3.2. The demand is satisfied by products which are made by two alternative technologies, i.e. product Q_i is made by technology i with i=1,2. Products Q_1 and Q_2 are perfect substitutes of each other. The total product costs are: (1) the non-materials costs of new products, $p_{qvj}Q_{vj}$ and the materials costs of a new product which are split up between the costs of virgin and recycled materials, $p_{mvi}M_{vi}$ and $p_{mri}M_{ri}$, (2) the costs of recycled products, $p_{qrj}Q_{rj}$ and (3) the costs of waste treatment, $p_{wdi}W_{di}$.

Figure II.3.2 A Two-Materials-One-Product Chain with two technologies.

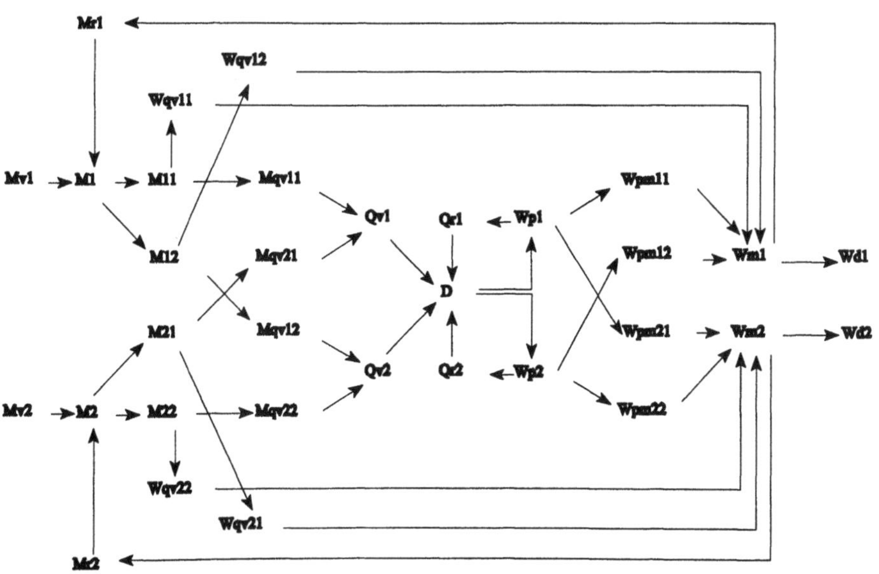

The objective function can then be formulated as:

$$\min_{c_{rmi},c_{rpj},M_{ij},Q_j} \Omega = \sum_{i=1}^{2} \sum_{j=1}^{2} p_{qvj}Q_{vj} + p_{qrj}Q_{rj} + p_{mvi}\, M_{vi} + p_{mri}\, M_{ri} + p_{wdi}W_{di} \quad (1)$$

The restrictions on the demand side, the production side and the waste and recycling side are given in Table Y and are analogous to those discussed in the previous two sections. After substituting the model conditions the objective function becomes:

$$\min_{c_{rmi},c_{rpj},M_{ij},Q_j} \Omega=\sum_{i=1}^{2} \sum_{j=1}^{2} p_{qvj}(1-c_{rpj})Q_j + p_{qrj}c_{rpj}Q_j +[(p_{mvi}+p_{wdi})(1-c_{rmi})+p_{mri}c_{rmi}]M \quad (2)$$

with the following conditions:

$$(1-c_{rpj})Q_j = f_j(M_{1j}, M_{2j}) \text{ for } j=1,2 \quad (3)$$

$$\sum_{j=1}^{2} Q_j = \overline{D} \quad (4)$$

$$0 \le c_{rmi}, c_{crpj} \le 1 \text{ for } i,j=1,2 \quad (5)$$

Table II.3.3 compares the two-technology with the one-technology case.

The decision variables are c_{rpj}, c_{rmi}, M_{ij}, Q_j for i,j=1,2. The optimal values for c_{rpj} and c_{rmi} are calculated using the LaGrange function, after which the optimal levels of the inputs of materials i for product j, M_{ij}, the quantity of products of type j made to satisfy the demand, Q_j, and the net costs are determined. In Kandelaars and van den Bergh (1996) it is shown that the optimal solutions for the c_{rmi} are extremes if the prices for new and recycled materials/reused products are not the same. Intuitively, the maximum amount of materials will be recycled if recycled materials are cheaper than new materials. The percentages of reused products (c_{rpj}), material inputs (M_{ij}), and the products (Q_j) to meet the demand depend on the production functions and the demand as indicated in the Appendix. Tables II.3.4 and II.3.5 give overviews of the MB conditions and the decision variables in the M-P chains described in Table Y.

Table II.3.3. Overview of models of the two-materials-one-product chain and the two technologies M-P chain.

	Two-materials-one-product chain (for i=1,2)	Two technologies (for i,j=1,2)
Demand side	$D = \overline{D}$ B	$D = \overline{D}$ C
	$Q_v+Q_r=D$	$Q_{vj}+Q_{rj}=Q_j$
	-	$\sum_{j=1}^{2} Q_j = DD$
Production side	E	$Q_{vj} = f_j(M_{1j}, M_{2j})$ F
	$M_i=M_{ri}+M_{vi}$	$M_i=M_{ri}+M_{vi}$
	-	$M_i = \sum_{j=1}^{2} M_{ij}$ G
Waste and recycling side	$M_i=W_{qvi}+M_{qvi}$	$M_i = \sum_{j=1}^{2} W_{qvij} + M_{qvij}$ H
	$W_p=D$	$W_{pj}=Q_j$
	$Q_r=c_{rp}W_p$	$Q_{rj}=c_{rpj}W_{pj}$
	$W_{pmi}=m_i(W_p-Q_{rj})$	$W_{pmij}=m_{ij}(W_{pj}-Q_{rj})$
	$m_i = \frac{M_i - W_{qvi}}{Q_v}$ I	$m_{ij} = \frac{M_{ij} - W_{qvij}}{Q_{vj}}$ J
	$W_{qvi}=g_i(M_i)$	$W_{qvij}=g_{ij}(M_{ij})$
	$W_{mi}=W_{qvi}+W_{pmi}$	$W_{mi} = \sum_{j=1}^{2} W_{qvij} + W_{pmij}$ K
	$M_{ri}=c_{rmi}W_{mi}$	$M_{ri}=c_{rmi}W_{mi}$
	$W_{di}=W_{mi}-M_{ri}$	$W_{di}=W_{mi}-M_{ri}$

Table II.3.4. Overview of the material balance conditions for the two M-P chains.

	Two-materials-one-product chain (for i=1,2)	Two technologies (for i,j=1,2)
Product level	$M_{qvi}=W_{pmi}$	$M_{qvij}=W_{pmij}$
Production level	$M_i=W_{mi}$	$M_i=W_{mi}$
Chain level	$M_{vi}=W_{di}$	$M_{vi}=W_{di}$

Table II.3.5. Overview of the decision variables of the two M-P chains.

	Two-materials-one-product chain (for i=1,2)	Two technologies (for i,j=1,2)
Recycling level	c_{rmi}, c_{rp}	c_{rmi}, c_{rpj}
Material inputs level	ratio between M_1 and M_2	ratio between M_{1j} and M_{2j}
Product level	-	ratio between Q_1 and Q_2

II.3.4 Results and interpretation

In M-P chain analysis economic, physical and environmental *indicators* are used to evaluate and compare various scenarios. Some economic indicators that are used are: the net costs; the total demand; allocation variable; government revenue of levies; average cost per product; changes in employment and trade; and, real output change of production sectors and household groups. M-P chain analysis also includes physical and environmental indicators, such as, the use of new or recycled materials, the waste of materials, the recycling percentage, raw materials depletion, water pollution, acidification, global warming potential, and energy use.

II.3.5 (Potential) links to other models

The driving force for the consumption of products is the desire for services. Therefore, materials, products and services should be studied together and simultaneously. The concept of an M-P chain includes an economic structure of connected material and product flows. With the concept of an M-P chain various analyses can be performed. In this chapter "M-P chain analysis" studies the allocation and economic processes of an M-P chain. This definition does not include LCA, MFA and I-O analysis, because these methods do not include allocation of materials and products, and economic processes. M-P chain analysis uses elements from MFA, physical I-O analysis and LCA, and combines those with an economic analysis. Table X.1 shows which economic and physical models are integrated in the applications. Note that an M-P chain is a concept to which various models may be applied.

The data on which an M-P chain analysis is based may be derived from a database on material and product flows through an economy. One of these database is FLUX (see Section II.2), which is used in the applied general equilibrium model. For the other models data of material flow models, such as FLUX, or LCA is used.

References

- Kandelaars, P.P.A.A.H. (1999. Economic Models of Material-Product Chains for Environmental Policy Analysis , Kluwer Academic Publishers, forthcoming.
- Opschoor, J.B. (1994). Chain management in environmental policy: analytical an evaluative concepts. In: J.B. Opschoor and R.K. Turner (eds.). Economic Incentives and Environmental Policies, Kluwer Academic Publishers, Dordrecht.

Legenda

Indices:

k-ore = z-ore for zinc ore, fe-ore for iron ore
i = z for zinc, p for PVC, gs for galvanised steel
j = zg for zinc gutters, pg for PVC gutters

Physical quantities (in kilograms):

$I_{k\text{-}ore}$ = inputs k-ore
M_i = total material input i
$M_{v,i}$ = virgin material input i
$M_{r,i}$ = recycled material input i
$W_{m,i}$ = waste material i
$W_{d,i}$ = dumped waste material i

Physical quantities (in functional units):

Q_j = gutter type j
Q_{fastj} = fastening-piece for gutter type j
D = demand for gutters

Prices (in guilders):

$p_{q,j}$ = price of one functional unit of a gutter of type j
$p_{mr,i}$ = price of one kilogram of recycled materials of type i
$p_{wd,gs}$ = price of the treatment of one kilogram of galvanised steel

Coefficients:

c_i = part of the waste material that is recycled
$x_{i,k\text{-}ore}$ = the amount of kilograms needed of k-ore to make one kilogram of material i
m_i = amount of kilograms of material i in gutter made of that material
b_j = amount of kilograms of galvanised steel needed for a fastening-piece of gutter j

Function

Ω = net costs of satisfying the demand

II.4 Heavy-metal balances of agricultural soils
Simon Moolenaar & Theo Lexmond

II.4.1 Introduction

Soil may be viewed as a natural resource because it is essential for production of food and fibre and for ecosystem functioning. Degradation of soil shows that soil is also a finite resource that needs sustainable management. Heavy-metal inputs to agricultural soils need not always be as small as possible since some metals are indispensable for life. According to Alloway (1990), there are three criteria for determining whether or not an element is essential i.e., the organism can neither grow nor complete its life cycle without an adequate supply of the element, the element cannot be wholly replaced by any other element, and the element has a direct influence on the organism and is involved in its metabolism. In contrast to cadmium and lead, copper and zinc are essential elements which may give rise to deficiency problems in plants and animals. Together with copper, zinc is primarily phytotoxic, so the concern about these metals is mainly directed at effects on crop yields and soil productivity. Together with cadmium, zinc may be considered as a mobile and bioavailable metal which may accumulate in crops and human diets (Kiekens, 1995). Food plants which tolerate relatively high concentrations of potentially hazardous metals create a greater health risk than those which are more sensitive. In general, it can be stated that food plants are more sensitive to copper and zinc than to lead and cadmium. Excessive uptake of both essential (copper, zinc) and non-essential (lead, cadmium) metals may result in adverse effects on soil biota, plants and, due to transfer via the food chain, on mammals, birds and humans.

If the content of essential heavy metals in soil is low, the supply to organisms will be inadequate and symptoms of deficiency (e.g., growth and yield reduction in the case of crops) may become manifest. In the deficiency range, a positive effect will result from increasing the soil content above the lower critical value as may be done by fertilizer additions. As the supply of the essential metals increases beyond this lower critical content, a level is reached where further increase does not have any effect e.g., on yield. This zone of luxury consumption is the optimal range. Increasing soil content beyond a certain upper critical content, induces adverse effects on soil biota (fauna and flora) and hence on biological activity (toxicity range).

The buffering capacity of soil with respect to soil contamination may be defined as its capacity to delay any (negative) effects of sustained additions of a contaminant because of inactivation. Inactivation is mainly achieved by effective bonding onto soil constituents or sometimes conversion into insoluble compounds. Beyond the upper critical content, soil is considered to be polluted as the buffering capacity is exceeded. The buffering capacity varies widely for different compounds and for different soils and reflects the soil's vulnerability (De Haan, 1996a).

The distinction between soil pollution and soil contamination reflects only a difference in degree of damage to the soil system. Any addition to soil of contaminants can be defined as soil contamination. Due to the soil's buffering capacity, it usually takes some time before negative effects of the contaminant's presence become apparent. Once this situation occurs the soil can be considered as polluted i.e., malfunctioning of the soil is apparent due to an abundant presence or availability of compounds (De Haan, 1996a). Since both essential and non-essential metals become toxic when critical contents are exceeded, excessive metal input to agricultural systems can be considered as a stress that potentially affects productivity and overall functioning of agro-ecosystems and the biodiversity of these systems (Ross, 1994).

Since accumulation of heavy metals in soil causes problems when certain soil contents are exceeded, control of heavy-metal fluxes is a prerequisite for sustainable agricultural production. To define strategies that ensure a sustainable cadmium (Cd), copper (Cu), lead (Pb), and zinc (Zn) management of agro-ecosystems, an analysis of the fluxes of these heavy metals in agriculture, their accumulation in agricultural soils, and associated risks, must be provided.

A soil protection policy with regard to heavy metals can be based on different principles. The state of the soil can be judged by studying relevant soil processes and adverse effects on important soil functions and soil organisms e.g., by a 'pathway analysis' (e.g., Chaney & Ryan, 1993). Such an approach calls for a thorough analysis of soil processes to reveal differences in vulnerability. Another perspective on soil protection is based on a more generic approach by using heavy-metal balance sheets as sustainability indicators (e.g., De Haan & Van der Zee, 1993). This 'balance approach' or 'flux approach' has proven to be very useful in soil fertility studies to discover the depletion of essential elements from soils (Frissel, 1978; Smaling, 1993). In the same way, the balance approach can be used in soil pollution research to discover the fate of both essential and non-essential elements in agricultural soils. This approach will be followed here.

A balance of heavy metals in soil relates the rates of input, accumulation, and output. The change in total heavy-metal content in soil depends on the input rate at the soil surface, the leaching rate at the lower boundary of the system and the removal rate by harvesting plants. Leaching and crop uptake of heavy metals are related to the metal concentration in the soil solution which depends on the labile heavy-metal content in soil and on sorption characteristics. The output rate coefficients depend on many chemical, physical, and biological properties. With projective dynamic balance calculations for the plough layer it is, in principle, possible to determine whether or not problems are likely to occur and, if so, in which compartment (soil, produce or groundwater) and on which time-scale (see also Chapter IV.3.4).

Agricultural systems receive heavy-metal inputs from various sources that differ in importance for different metals, different systems and different soils. Input flows are caused by primary (i.e., economic) and secondary (i.e., inputs via the environment) sources. Primary sources are part of the material flows that come from industry (e.g., mineral fertilizers, feed), trade (e.g., straw, animals, manure), waste management (compost), and waste water management (sewage sludge). Within the group of primary

sources a distinction can be made between intentional inputs (e.g., Cu-compounds used as pesticide or fertilizer) and non-functional inputs (e.g., Cu as a constituent of soil amendments, like manures). Since primary sources consist of means of production, i.e., purchased and self-processed materials, they may be controlled to some extent. Secondary sources cause a non-functional and uncontrollable input into agro-ecosystems, like in the case of atmospheric dry and wet deposition of heavy metals and sedimentation after inundation in areas that are regularly flooded. The distinction between primary and secondary sources can be used for taking reduction measures that are aimed at the right target group. Output proceeds via marketing produce (trade) and via losses to the environment. The output via produce depends on the farming system and it will reach the consumers through the trade and retailing chain. Losses to the environment are mainly due
to processes like leaching, wind erosion and surface runoff. Figure II.4.1 provides a schematic overview of agricultural soil with its inputs and outputs.

Figure II.4.1 Inputs and outputs of agricultural soils

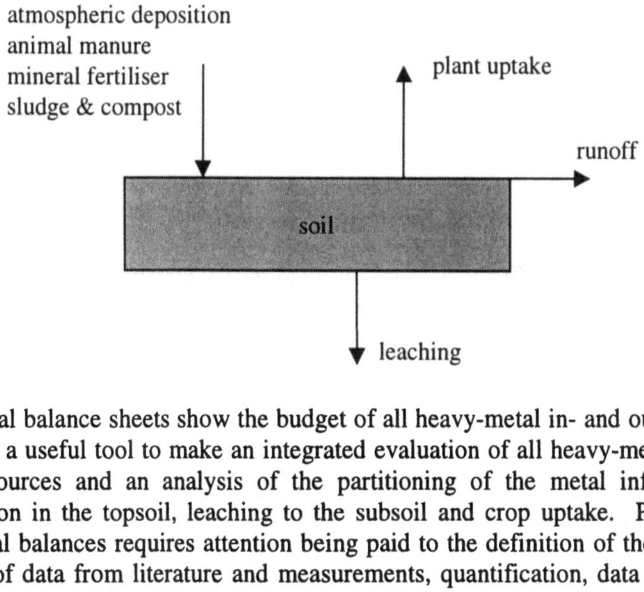

Heavy-metal balance sheets show the budget of all heavy-metal in- and outputs, which make them a useful tool to make an integrated evaluation of all heavy-metal inputs by different sources and an analysis of the partitioning of the metal influx between accumulation in the topsoil, leaching to the subsoil and crop uptake. Proper use of heavy-metal balances requires attention being paid to the definition of the system, the reliability of data from literature and measurements, quantification, data presentation, and interpretation of the balance in view of sustainability (Moolenaar & Lexmond, 1999).

II.4.2 Goal and Scope

Heavy-metal flows of agricultural systems can be analyzed on the field scale, the farm scale, or the (supra-) national scale.

Analysis on the (supra-) national scale

The national balance is calculated by subtracting all heavy-metal flows leaving agriculture from all the heavy-metal flows entering agriculture. Thus, the total net input of heavy metals to the (agricultural) soil is calculated for agriculture as a whole and one gets an overview of this 'average burden' on the 'average agricultu-ral soil' by applying national statistics on feedstuffs, mineral fertilizers, animal manure, agricultural products (milk, meat, crops), etc. Using annual sales, mean concentrations (e.g., in crops or fertilizers), mean application rates, and mean yields per crop, the mean annual loads in a certain region can be calculated. This method allows for tracing down the most important heavy-metal flows and bottlenecks on a regional to (supra-) national scale. Clearly, analyses at the national level do not give due attention to relevant processes on a smaller scale, because the local and site specific aspects with regard to soil characteristics (e.g., adsorption capacity, hydrological properties) and soil use (e.g., inputs and crop rotation) are averaged out.

Analysis on the farm scale

The farm-gate balance shows the characteristic flows and processes of the farm as a whole. 'Farm-gate' refers to flows that can be measured when entering or leaving the farm 'gate', so this term is the agricultural equivalent to a facility-level mass balance in manufacturing where the term 'factory-gate' is used. Since large differences exist for different metals between and within farming systems, the farm-gate balance offers a way to finetuning heavy-metal management directly at the farm level. The internal flows between the farm compartments (i.e. subdivisions of the system like crop, animals, soil) are not accounted for in a farm-gate balance. These compartments have different input and output flows, which differ largely between farming systems. Inputs of the animal compartment consists of feed and litter (used indoors), consumption of harvested crop (roughage), and grazing of forage. Eggs, milk meat and manure are the outputs. The inputs of plants are related to uptake and atmospheric deposition, while the output of this compartment is the consumption by grazing or harvesting of the plants themselves as primary products (Moolenaar 1999).

Analysis on the field scale

The field-scale balance shows the heavy-metal balance for the soil compartment or the plough layer of individual fields. A balance of heavy metals on the field scale relates the rates of accumulation, inputs and outputs. The input and output rates are a function of soil and management characteristics, which can be derived for individual fields (Moolenaar 1998).

In this chapter and in chapter III.3.3, the farm scale and field scale analyses are described. In view of the diversity of the relations with the external surroundings, agricultural systems may be typified as open systems. The boundary of a system is the imaginary line seperating what is considered to be inside the system and what is considered to be outside. The single farm has, in principle, easily recognizable boundaries. The agricultural activities at the single farm (or field) level consist of animal production and/or crop production. The system is open to the atmosphere and the upper side of the farming systems is formed by crops and buildings. Its lateral borders are defined by the farm gate (property boundaries) or by surface waters.

The appropriate system boundaries of heavy-metal flows and cycles may be defined for different spatial and temporal scales. With regard to the temporal boundary, it is impossible to choose a time-base to suit the rates of all processes. Thus, the chosen

timebase will always be somewhat arbitrary like e.g., one year (i.e., growing season) or one crop rotation. The interactions between current, preceding and subsequent crops justifies the drawing of system boundaries around a crop rotation rather than around a particular crop.
With regard to the spatial (physical or topographical) boundary, it may be that the same entities can be considered as being both part of the economic system and part of the environment since the different functions of the system can hardly be separated with respect to the economic and the ecological subsystems. For example, the topsoil is defined as the upper part of the soil (ca. 0.3 m) that is intensively used by man during agricultural activities. The plough layer (the soil that is turned by a plough) can therefore be included within the system boundary since it is an integral part of the agricultural production system (economic function). Moreover, the farming activities change the soil's quality which in turn influences the output. For this reason, Cowell (1998) argues that soil should be considered an ancillary i.e., material that contributes to maintenance of processes but that is not intended to enter the product. Soil should thus be included within the system boundary because its quantity and quality are closely linked with the farming activities taking place on the land. Indeed, the soil can also be regarded as a non-renewable resource that is "processed" and "formed" by practices such as fertilization and tillage. This view contrasts with the opinion of researchers who regard the plough layer as being part of the environment (related to ecological functions). In that case, it is separated from the farming system and all flows into the plough layer should then be considered as (output) flows to the environment. Here, the 'default' choice is that the plough layer belongs to the environment and thus heavy-metal emission to the topsoil is regarded as an emission to the environment.
In the following section we discuss the modeling of heavy-metal flows on the field scale taking into account a detailed analysis of soil contents, crop uptake and leaching out of the topsoil.

II.4.3 Modelling principles

A heavy-metal balance equation for the topsoil for the plough layer relates the rates of change in heavy-metal content, input and output according to the mass conservation principle and is given by

$$\Delta G = I - O$$

The change in total heavy-metal content in the plough layer (ΔG) is the balance of the input (I) at the soil surface and the output (O) by leaching out of the plough layer and by removal in harvested products. Losses to the atmosphere in the case of volatile metals and losses by (wind and water) erosion may be incorporated in the balance equation if appropriate.

Static balance
In a "static" balance (SB), a record is kept of the input and output flows for one year or one crop rotation. In a SB approach, the output flows are assumed not to be related to the (total) metal content in the soil. The change in heavy-metal content in the plough layer is therefore the result of the net difference between input rate and (constant)

output rates. Because a SB does not regard the dependency between soil content and output flows, it cannot realistically simulate the heavy-metal soil content in time. A static balance is comparable to a black box model which serves to find relations-hips between input and output of a system, without knowing the system's structure and behavior. In a black box model, the input-output relationships are determined by some transformation function. In the case of soil, this transformation function is not constant due to changes in the soil's buffering capacity etc. The influence of system's properties on the output in time is determined by the system's and compound's behavior. The state of a system at a moment in time is the set of relevant properties which that system has at that time. This state needs to be known in order to predict, with a given input, the output (deterministic) or the probability of a certain output (stochastic) in the course of time. This is attempted in dynamic modeling (Moolenaar & Lexmond, 1999).

Dynamic balance

For simulating the long-term behavior in time, a 'dynamic' balance (DB) may be calculated in which the relationship between soil content (G: g m^{-3}) and output flows in time are explicitly included. The DB equation can thus be derived in terms of heavy-metal content by relating the output rates to the total metal content in soil:

$$\mathrm{d}G/\mathrm{d}t = A - L - U$$

In this equation, the change in total heavy-metal content in soil ($\mathrm{d}G/\mathrm{d}t$) depends on the input rate at the soil surface (A), the leaching rate at the lower boundary of the system (L) and the removal rate by crops (U), which are all expressed in (g m^{-3} yr^{-1}). Because the balance calculations usually involve rather large time scales, intra-seasonal variations in crop uptake, leaching and plough layer composition are reduced by averaging out over many growing seasons. In an annually ploughed field, the plough layer may be considered to be homogeneous (*i.e.*, perfectly mixed).

Leaching rate (L):

The total amount of heavy-metals in soil (G) is the sum of fixed, adsorbed, and dissolved heavy metals. The fixed fraction of heavy metals cannot be released into the soil solution while the adsorbed amount participates in the equilibrium reactions between the soil solid and solute phases. By measuring adsorption isotherms, the relationship between the heavy-metal concentration in solution (c: g m^{-3}) and the adsorbed amount (q: g kg^{-1}) can be derived assuming that equilibrium is instantaneous and that no desorption hysteresis occurs. Adsorption of a heavy metal can for instance be modeled with the Freundlich sorption equation, given by:

$$q = k_f c^n$$

where k_f (g^{1-n} $(m^3)^n$ kg^{-1}) and n (-) are constants. Using the Freundlich equation, the labile content G_l (i.e., the sum of the adsorbed amount in the solid phase and the dissolved amount in the soil solution) is expressed as:

$$G_l = \theta c + \rho k_f c^n$$

Here, θ is the volumetric water content of the soil ($m^3\ m^{-3}$) and ρ is the soil dry bulk density (kg m^{-3}). Commonly, the distribution ratio defined by

$$r_d = \rho q/\theta c$$

is large. Hence, θc can be neglected to obtain an approximation for the relationship between c and G_l:

$$c = [G_l/\rho\, k_f]^{1/n}$$

The leaching rate (L) is the product of precipitation surplus and heavy-metal concentration in solution (c). Consequently, the leaching rate from the plough layer can be related to the labile fraction of the total soil content (G_l: g.m^{-3}):

$$L = P_s c/\, l_p = \theta v c/\, l_p = \theta v[G_l/\rho\, k_f]^{1/n}/\, l_p$$

where P_s is the precipitation surplus (m yr^{-1}), v is the pore water velocity (m yr^{-1}), and l_p is the plough layer thickness (m).

Plant uptake rate (U):

Although there are many soil and plant factors that influence heavy-metal availability to plants, crop uptake is expressed here according to the relationship

$$U = BG_l$$

(cf. Kuboi et al., 1986). Here, the plant uptake rate coefficient, B (yr^{-1}), can be related to the crop yield (Y: kg m^{-2} yr^{-1} DW), the metal content in the crop (cp: (mg kg^{-1} DW), and the labile soil content according to:

$$B = Yc_p/l_p G_l$$

In this equation, it is assumed that metal uptake is limited to the plough layer. Other factors that influence the burdening of plants with heavy metals (e.g., interception) could be incorporated as well.

The dynamic balance of the topsoil:

A differential equation represents continuous changes of state with time and by substituting the eqautions for leaching (L) and plant uptake (U) in the general dynamic balance equation, a differential equation develops in terms of (labile) heavy-metal content in the topsoil:

$$dG/dt = A - U - L = A - BG_l - C[G_l]^{1/n}$$

in which the leaching rate parameter (C) is given by:

$$C = \theta v[1/\rho\, k_f]^{1/n}/\, l_p$$

Analytical solutions are available for particular values of n, whereas for other cases the solution of this equation must be approximated numerically (Boekhold & Van der Zee, 1991; Moolenaar, 1998).
The rate parameters (*A, B, C*) are lumped i.e., within certain spatial and temporal limits, they are considered as constants. These rate parameters are a function of soil and management characteristics and they can be derived from measurements or estimations for any agro-ecosystem. With this dynamic balance of the topsoil, soil contents can be calculated as a function of time, revealing if problems are likely to occur and, if so, for which element, in which compartment (soil, produce or groundwater) and on which time-scale (see also Chapter IV.3.4 and Moolenaar et al., 1997a).

Soil-bound heavy-metal flows:
The role of 'soil-bound' heavy-metal flows with regard to calculating heavy-metal balances may become an important issue in SFA studies. The substances which are studied (i.e., heavy metals in this case) are mostly part of materials which serve as their carriers. In order to study certain substances it is thus necessary to carry out a material flow analysis (MFA) at the same time. The carriers or materials which are studied in MFA may be very important for the resulting SFA. For example, Cd may be 'carried' by P-fertilizers and by refuse compost. Due to the great differences in the matrices of these materials, these different types of carriers influence the form and rate of Cd accumulation in soil.
The losses through 'soil-bound' heavy-metal flows like dirt tare, runoff and wind erosion may be a significant burden to other environmental compartments. In the SB and DB approaches, the heavy metals that are leaving the system as 'soil-bound' flows are regarded as regular output flows. Discriminating between output flows that are soil-bound and output flows that are not complicates the calculations considerably, because the matrix of heavy-metal carriers is to be taken into account properly. In order to solve these complications, Moolenaar et al. (1997b) developed the 'dynamic soil composition' balance (DSCB) approach.

Dynamic soil composition balance
Changing heavy-metal content in the plough layer is the result of the net difference between input and output flows per unit time. Existing balance approaches lack an analysis of the effect of soil amendments on soil composition. However, soil composition determines soil bulk density and plough layer weight and hence the resulting change in heavy-metal content. The dynamic soil composition balance approach accounts for changing soil composition by calculating mass balances of both heavy metals and main soil constituents.

An example in which the addition of constituents other than heavy metals has to be considered is the recycling of compost produced from source-separated organic (SSO) household waste, i.e., kitchen waste, residual food and yard trimmings. In the Netherlands, SSO-waste is composted. Agriculture is a potential market for the SSO-compost to improve physical soil characteristics (organic matter status, soil structure and water holding capacity). SSO-compost also has some value as a fertilizer and may raise soil pH. Addition of SSO-compost may cause heavy-metal accumulation because of contamination of compost by heavy metals (TCB, 1991; Fricke & Vogtmann, 1994;

Richard & Woodburry, 1994; Deportes et al., 1995). A Dutch General Administrative Order (GAO) of 1991 sets limit values for the maximum heavy-metal contents in compost and in the receiving soil, and for the maximum application rates. Dutch SSO-compost contains approximately 70% dry matter, comprising 30% (expressed on dry weight: DW) organic matter and 70% (DW) soil minerals. In the SB- and DB-approaches, the heavy-metal content changes linearly with compost additions without taking into account the soil that is added at the same time. The DSCB-approach explicitly accounts for the gradual changes in soil composition in the plough layer as a result of the inputs of soil minerals as well as organic matter. Thus, the changes in heavy-metal contents are related to the changes in solid phase composition, such as the clay, non-clay, and organic matter fractions, as these constituents are chemically relevant. The total content of heavy metals is calculated with the mass balances for heavy metals, organic matter and other soil constituents (clay, sand and silt) in the plough layer. These fractions determine the soil bulk density and hence the plough layer weight. The organic matter dynamics are taken into account in a organic matter balance. In the DSCB calculation the organic matter dynamics are modeled according to the 'apparent initial age' concept of Janssen (1984), because this model has proved a useful and accurate tool for simulating organic matter dynamics.
Besides the input of soil material and organic matter dynamics, changes in the soil surface level after repeated (compost) additions are taken into account in the DSCB-approach, because applications of organic amendments may result in a small yet significant rise of the soil surface level. Corrections based on the soil surface level influence the mass balances directly, because if the soil surface level rises slowly, the plough layer is kept at a constant thickness. As a result, a small part of the heavy metals is 'lost' from the plough layer to the deeper soil layers as mixing by ploughing is limited to e.g., 0.3 m. Therefore, the DSCB-model has a correction factor for the 'increasing' soil level and constant plough layer thickness. Some processes, like decomposition of organic matter, losses through dirt tare, runoff and wind erosion can reduce this increase in level or even result in a net decrease of the soil level. These processes are accounted for by using the correction factor.

The calculation of the resulting heavy-metal contents according to the DSCB approach is thus based on the mass balance of all relevant soil components in the plough layer, including organic matter dynamics and changing soil level. While conventional balance approaches simply assume that accumulation is proportionally related to the amount of heavy metals applied, DSCB-calculations distinguish the effect of the matrix of soil amendments on the resulting accumulation in the plough layer. If only mineral fertilizers are used, a quite different situation may occur compared with the use of organic amendments on long time scales. In the case of applying soil amendments with high soil and organic matter fractions, it is therefore better to account for the composition of both the soil amendment and of the soil when calculating heavy-metal balances. The DSCB-approach is thus especially relevant if the input consists of animal manure, other organic fertilizers or organic soil conditioners like SSO-compost. Moolenaar et al. (1997b) showed that the distinction between output flows that are 'free' (e.g., leaching and plant uptake) and those that are 'bound' (e.g., dirt tare and erosion) may be very important for the outcome of heavy-metal balance calculations. With the DSCB-approach these soil-bound heavy-metal flows can be accounted for correctly. The practical implications of this new approach for MFA/SFA studies and for setting limits with regard to both heavy-metal input flows that are 'free' (deposition, mineral fertilizers) as well as input flows that are 'bound' to a soil matrix (organic amendments) should therefore be assessed.

II.4.4 Results and Interpretation

The difference between input flows (*I*) and output flows (*O*) determines the resulting accumulation (expressed as *I-O* or *I/O*). Although accumulation indicates that the system is off balance, accumulation per se as a primary effect only reveals that there is an absolute increase in the soil's stock (*S*). Relating accumulation to the total amount of metals in the soil ([*I-O*]/*S*) already gives some more insight in the potential risk for environmental problems in the future since it shows the relative (or percentual) stock increase (*i.e.*, ΔS). In order to calculate the resulting change in soil content (ΔG) from the stock increase, it is necessary to know the soil bulk density and the thickness of the mixed layer as well. These (location specific) soil parameters may not be constant in time.

An analysis of heavy-metal flows and resulting (absolute or relative) accumulation should be followed by an analysis of what happens in the soil in order to predict the effects of accumulation on production and human and ecosystem health (*i.e.*, 'environmental quality'). Hence, indicators that relate effects on different environmental compartments require consideration of the relevant physico-chemical processes in a long-term perspective. For instance, leaching causes the net accumulation to decrease but high leaching rates may threaten groundwater quality in turn. A rule of thumb should therefore be applied to both essential and non-essential elements. For essential elements, input should equal output (at an optimum level of supply). For non-essential elements, the input should be smaller than or at the most equal to the permissible output (i.e., output that does not exceed quality standards for soil, ecosystems, crop and groundwater).

If an element is naturally abundant (*e.g.*, iron and aluminum) and accumulates in a chemical form of similar solubility to its natural compounds, accumulation may proceed without any appreciable effect on the system itself, its produce or on the environment. However, accumulation of Cd, Cu, Pb, and Zn mostly involves a steady increase in activity and/or mobility in the soil. The rate of this increase depends on the soil's buffering capacity and on the actual surplus on the balance sheet. Accumulation of these elements thus leads to an increase in element flows between the system compartments (resulting in increasing contents in produce) and from the system to its surrounding environment (Van Riemsdijk *et al.*, 1987). Since cycling features may be modified, both mass balances and transfer rates should be considered. The dynamics of elements' pathways and transfers between the different pools are determined by their solubility in water, their degree of chemical reactivity, and by the physical and chemical environment. This results in different cycling characteristics for different elements (Frissel, 1978) and thus a system may be in different states of balance for different elements at the same time (Van Riemsdijk *et al.*, 1987).
Since balance studies try to avoid to account for the above mentioned processes in a detailed manner, another way to account for 'risks' or effects is to make use of existing

standards that are related to the toxicity of the metals. If the critical soil content is expressed as G_{crit}, the rate of reaching this critical value equals $\Delta G/[G_{crit}-G]$. This kind of calculations may be carried out *e.g.*, by using standards of the European Community (C.E.C., 1986) and the United States EPA (USEPA, 1993).

Apart from the problem that these standards do not always have a sound scientific basis (as is discussed in McGrath et al., 1994), another setback of this approach is that the resulting rates are usually based on static balance calculations (using constant output rates), whereas in reality the output flows (e.g. by leaching and crop offtake) are dynamically related to the soil content in the case of Cd, Cu, Pb, and Zn. Therefore, Moolenaar *et al.* (1997a) derived indicators of sustainable heavy-metal management in agro-ecosystems from a dynamic soil balance. These indicators provide insight in the relative importance of different heavy-metal flows and allow priority assessment of protection measurements and quantification of the gains of management options that aim at preventing quality standards for soil, crop, and groundwater from being transgressed.

Sustainability indicators

A dynamic balance of heavy metals in soil relates the rates of accumulation, input and output and can be given by:

$$dG/dt = A - U - L = A - BG_l - C[G_l]^{1/n}$$

In this equation, the change in total heavy-metal content in soil (dG/dt: g m^{-3} yr^{-1}) depends on the labile fraction of the total soil content (G_l: g m^{-3}), the input rate at the soil surface (A: g m^{-3} yr^{-1}), the leaching rate at the lower boundary of the system (L: g m^{-3} yr^{-1}), and the removal rate by harvesting plants (U: g m^{-3} yr^{-1}). With this dynamic balance equation, soil contents can be calculated as a function of time, revealing if problems are likely to occur and, if so, for which element, in which compartment (produce or groundwater) and on which time-scale.

In reality, the parameters related to plant uptake and leaching depend on many chemical, physical and biological properties of the soil-plant-system. The combination of a huge variety of soil properties and these chemical, physical and biological conditions makes the development of general rules for quantitative evaluation of soil quality a difficult task. Hence, the assessment of the heavy-metal balance for many different situations is often impossible as notably the plant uptake and leaching rate parameters may be difficult to quantify. Furthermore, a balance assessment at a larger scale may lead to wrong conclusions locally. Fortunately, characteristic numbers can be derived from the dynamic balance that either do not suffer from these shortcomings or that enable us to prioritize which limited data should be experimentally assessed in case commonly available data suggest that problems with regard to heavy metals should be anticipated. Using the input (A) and output (B, C) rate parameters, a quantitative evaluation of specific local situations and of characteristic 'general situations' can be carried out with the help of sustainability indicators which are based on existing or proposed quality standards. They serve as indicators for adverse effects on the soil and related compartments and they are useful in view of the often limited availability of (reliable) data with respect to the soil-plant-system and input parameters that are needed to characterize an agro-ecosystem.

Two cases, *i.e.*, a linear (n=1) and a non-linear (n=1/2) balance equation will be studied for illustrative purposes. The values of the input rate parameter (A), plant uptake rate coefficient (B), and leaching rate parameter (C: yr^{-1} if n=1 and $m^3\ g^{-1}\ yr^{-1}$ if n=1/2) are assumed to be constants in both cases.

The linear balance

For n=1, the dynamic balance equation becomes:

$$dG/dt = A - (B + C)G_l$$

It can easily be seen that $(B+C)$ is equal to the 'elimination rate constant'. Integrating this balance, yields, for initial content G_0:

$$G(t) = \frac{A(1\, e^{(B+C)t}) + (B+C)G_0\, e^{(B+C)t}}{B+C}$$

which reduces to a more simple form for negligible initial heavy-metal contents in soil ($G = 0$ at $t = 0$). The value of G at steady state (ss) is given by:

$$G(ss) = \frac{A}{B+C}$$

The non-linear balance

Often, non-linear adsorption isotherms are observed for heavy metals (De Haan *et al.*, 1987). If the value of n equals ½, implying strong non-linear adsorption behavior, the balance equation becomes:

$$\frac{dG}{dt} = A - BG_l - CG_l^2$$

Integrating this balance yields an analytical solution:

$$G(t) = \frac{2A(\, e^{t\sqrt{D}}\, 1) - (B - \sqrt{D}\,)G_0\, e^{t\sqrt{D}} + (B + sqrtD\,)G_0}{2\,CG_0(\, e^{t\sqrt{D}} - 1) - (B - \sqrt{D}\,) + (B + \sqrt{D}\,)e^{t\sqrt{D}}}$$

with

$$D = B^2 + 4AC$$

which reduces to:

$$G(t) = \frac{2A(\,e^{t\sqrt{D}}\ 1)}{[(B + \sqrt{D}\,)e^{t\sqrt{D}} - B + \sqrt{D}\,]}$$

for negligible initial heavy-metal contents in soil ($G = 0$ at $t = 0$). For the non-linear balance, the value of G at steady state is given by:

$$G(ss) = \frac{B - \sqrt{D}}{2C}$$

The discrepancy factor
At steady state, the accumulation rate equals zero and therefore:

$$A = B\,G_l + C\,G_l^{1/n}$$

This means that the input rate equals the sum of the output rates. If we replace the output rates by the maximum acceptable output rates based on existing quality standards for crop and groundwater, we can define the *discrepancy factor* (F_d) for the soil compartment. This yields:

$$F_d = \frac{A}{U_c + L_c}$$

Instead of using the quality standards for crops (available in the case of essential metals), phytotoxic limit values might be used as well (available in the case of non-essential metals) to determine the total critical output rate. Since the discrepancy factor compares the input rate with the total acceptable (or critical) output rate it is directly related with inputs (A) from agricultural and non-agricultural sources and with standards for acceptable crop quality (maximum acceptable removal rate by harvest: U_c) and groundwater quality (maximum acceptable leaching rate: L_c). If the discrepancy factor exceeds 1, problems are expected to occur since the input rate exceeds the sum of allowable output rates. By comparing the discrepancy factors for different metals we can assess which heavy metal will eventually lead to the largest violation of (groundwater or crop) standards, *i.e.*, which metal is relatively most 'abundant'. So, the power of F_d is that it allows for prioritization between different metals (De Haan & Van der Zee, 1993).

The value of F_d may underestimate the real discrepancy between input and acceptable output since it uses the summation of U_c and L_c. In practice, one of these two removal rates determines which heavy-metal input is still acceptable. Moreover, the discrepancy factor does not take into account any standards for soil ecology. Therefore, the value of F_d serves as a first indicator of potential problems only.

The critical sustainability factor

The discrepancy factor does not reveal whether problems are due with regard to soil, crop or groundwater quality. If only limited data regarding water flow, heavy-metal sorption, mobility and bioavailability are available, a more advanced assessment is feasible already. The ranking of the threat to the different compartments at steady state depends on which limit is exceeded most and can be assessed with the *sustainability factors* for ecology (F_e), crop uptake (F_u), and soil solution or leaching (F_s). Thus, the most threatened object may be identified by comparing the steady-state values of the soil content, the crop uptake rate and the leaching rate with the corresponding critical values. This is shown by the *critical* sustainability factor (F_c), given by:

$$F_c = \mathrm{MAX}(F_e, F_u, F_s) = \mathrm{MAX}\left(\frac{G_{SS}}{G_e}, \frac{BG_{SS}}{U_c}, \frac{C\,G_{SS}^{\;1/n}}{L_c}\right)$$

which may be different for different heavy metals and with G_e as an ecological soil quality standard (*e.g.*, Van Straalen & Denneman, 1989). This standard could be used to include phytotoxicity to cultivated crops as well.

The critical sustainability time

Whereas the above characteristic numbers do not yield information on the time when standards will be violated, for each threatened function expressions for these *sustainability times* can be derived from the dynamic balance (t_e: time at which the ecological quality standard is exceeded; t_u: time at which the crop quality limit is exceeded; t_s: time at which the groundwater limit is exceeded). The *critical* sustainability time (t_c) identifies the compartment for which the quality standard (G_e, U_c or L_c), if exceeded, is exceeded first and is thus defined as:

$$t_c = \mathrm{MIN}(t_e, t_u, t_s)$$

A smaller value for U_c, L_c, or G_e, results in a smaller value for the respective sustainability times. In the linear case, t_e is given by:

$$t_e = \frac{\ln \dfrac{[A(B+C)G_0]}{[A(B+C)G_e]}}{B+C}$$

The expressions for t_u and t_s can be found by replacing G_e by U_c/B and L_c/C, respectively. The same expressions can be used in the case of negligible initial heavy-metal contents by setting $G_0 = 0$. It is assumed here that G_0 is smaller than the critical G values and consequently that A (input rate) exceeds the sum of B and C during the accumulating phase. These equations show that the largest sustainability factor has the smallest sustainability time for the linear model. Hence, the compartment that is threatened most at steady state is also threatened first.
In the non-linear case, t_e is given by:

$$t_e = \frac{\ln\left[\dfrac{2A-(B+\sqrt{D})G_0-(B-\sqrt{D}+2CG_0)G_e}{2A-(B-\sqrt{D})G_0-(B+\sqrt{D}+2CG_0)G_e}\right]}{\sqrt{D}}$$

The equations for t_u and t_s can be found by replacing G_e by U_c/B and $(L_c/C)^{1/2}$, respectively. Observe that in the non-linear case $(L_c/C)^{1/2}$ does not equal the reciprocal of the sustainability factor as in all other cases. Due to non-linearity of the adsorption equation, the largest critical sustainability factor does not necessarily correspond with the smallest sustainability time.
The advantage of using sustainability times and sustainability factors is that they are useful in comparing different systems. The calculation of these sustainability indices allows for a quick and efficient assessment of the information that is most relevant. Moreover, the results comply with the often limited data availability: a more specific analysis requires high quality input data. With the current generally available data it seems that the assumption of linearity is not the main constraint with respect to how realistic the approach is. As long as the standards for soil, groundwater and product quality are at best indicative of possible effects, relative criteria for sustainability as presented here may serve well for the purpose of classification and prioritization.

The rate parameters (A, B and C) largely determine the long-term soil contents and the value of the sustainability indices. The value of the soil content above which the critical crop contents will be exceeded (G_u: g m^{-3}) is very sensitive to the choice of the acceptable crop quality and to the value of the plant uptake rate parameter.
The long-term behavior of heavy metals can be simulated using the mean area-weighted values of the input and output rate parameters for each type of agro-ecosystem. The value of A (input rate) is based on atmospheric deposition and fertilizer applications. The value of U_c is the product of the yield (Y) and the quality standards (or phytotoxic limit values) of the crops involved. The value of L_c is the product of the quality standard for groundwater and the groundwater recharge rate. These values are related with corresponding (critical) soil contents for crop uptake and leaching (G_u and

G_s, respectively). The groundwater quality standard may be based on the Dutch target value for heavy-metal concentrations in groundwater.

II.4.5 Links to other models

Dynamic heavy-metal balances of agricultural soils may be linked to other models in two distinctive ways. One way is to couple the D[SC]B model with well defined metal speciation models in order to get more detailed insight in the behaviour and fate of different metal species. This was done by Moolenaar et al. (1998) who studied the effect of increasing (solid and dissolved) organic matter and decreasing pH as a result of changing land use from agriculture to forestry. The speciation of Cu and the consequent multi-phase competition (Cu complexation with organic matter) was taken into account in dynamic Cu balances of the topsoil. The results show that copper speciation changes dramatically with far reaching consequences for mobility and bioavailability.
Another way is to use D[SC]B modelling as part of a 'research train' as described in chapter IV.3.4. In that case it enables a powerful analysis in combination with more generic models such as Dynabox which is described in II.5.

References:

- Alloway, B.J. (ed.) (1990). Heavy metals in soils. Blackie, Glasgow, 339 pp.
- Boekhold, A.E. & S.E.A.T.M. van der Zee (1991). Long term effects of soil heterogeneity on cadmium behaviour in soil. Journal of Contaminant Hydrology 7: 371-390.
- Commission of the European Communities (C.E.C.) (1986). Council Directive (86/278/ EEC) on the protection of the environment, and in particular of the soil, when sewage sludge is used in agriculture. Official Journal of the European Community L181 (Annex 1a), pp. 6-12.
- Chaney, R.L. & J.A. Ryan (1993). Heavy metals and toxic organic pollutants in MSW-com-posts: Research results on phytoavailability, bioavailability, fate, etc. pp. 451-506. In: H.A.J. Hoitink & H.M. Keener (eds.): Science and engineering of composting: Design, environmental, microbiological and utilization aspects, Renaissance, Worthington OH, pp 728.
- Cowell, S.J. (1998). Environmental Life Cycle Assessment of Agricultural Systems: Integration Into Decision-Making. Ph.D. thesis. Centre for Environmental Strategy, University of Surrey, Guildford.
- De Haan, F.A.M., S.E.A.T.M. van der Zee & W.H. van Riemsdijk (1987). The role of soil chemistry and soil physics in protecting soil quality: variability of sorption and transport of cadmium as an example. Netherlands Journal of Agricultural Science 35: 347-359.
- De Haan, F.A.M. & S.E.A.T.M. van der Zee (1993). Compost regulations in The Netherlands in view of sustainable soil use. In: H.A.J. Hoitink & H.M. Keener (eds.): Science and engineering of composting: Design, environmental, microbiological and utilization aspects. Renaissance, Worthington OH, pp. 507-522.

- De Haan, F.A.M. (1996a). Soil quality evaluation. In: F.A.M. de Haan & M.I. Visser-Reyneveld (eds.): Soil Pollution and Soil Protection. Wageningen Agricultural University and International Training Centre (PHLO), Wageningen, pp. 1-17.
- Deportes, I., J.L. Benoit-Guyod & D. Zmirou (1995). Hazard to man and the environment posed by the use of urban waste compost: A review. Science of the Total Environment 172: 197-222.
- Fricke, K. & H. Vogtmann (1994). Compost quality: Physical characteristics, nutrient content, heavy metals and organic chemicals. Toxicological and Environmental Chemistry 43: 95-114.
- Frissel, M.J. (ed.) (1978). Cycling of mineral nutrients in agricultural ecosystems. Agro-ecosystems 4: 1-354.
- GAO (1991). Besluit kwaliteit en gebruik overige organische meststoffen. Staatsblad 1991: 613.
- Janssen, B.H. (1984). A simple method for calculating decomposition and accumulation of "young" soil organic matter. Plant and Soil 76: 297-304.
- Kuboi, T., A. Noguchi & J. Yazaki (1986). Family-dependent cadmium accumulation characteristics in higher plants. Plant and Soil 92: 405-415.
- Kiekens, L. (1995). Zinc. In: B.J. Alloway (ed.); Heavy metals in soils. Blackie, Glasgow, pp. 284-305.
- McGrath, S.P., A.C. Chang, A.L. Page & E. Witter (1994). Land application of sewage sludge: Scientific perspectives of heavy metal loading limits in Europe and the United States. Environmental Review 2: 108-118.
- Moolenaar, S.W., S.E.A.T.M. van der Zee & Th.M. Lexmond (1997a). Indicators of the sustainability of heavy metal management in agro-ecosystems. The Science of the Total Environment 201: 155-169.
- Moolenaar, S.W., Th.M. Lexmond & S.E.A.T.M. van der Zee (1997b). Calculating heavy metal accumulation in soil: A comparison of methods illustrated by a case-study on compost application. Agriculture, Ecosystems and Environment 66: 71-82.
- Moolenaar, S.W. (1998). Sustainable Management of Heavy Metals in Agro-ecosystems. Ph.D. thesis, Wageningen Agricultural University, Wageningen, 191 p. ISBN 90-5485-835-4.
- Moolenaar, S.W., E.J.M. Temminghoff & F.A.M. de Haan (1998). Modeling dynamic copper balances for a contaminated sandy soil following land use change from agriculture to forestry. Environmental Pollution 103: 117-125.
- Moolenaar, S.W. (1999). Heavy-metal balances. II: Management of cadmium, copper, lead and zinc in European agro-ecosystems. Journal of Industrial Ecology 3 (1), in press.
- Moolenaar, S.W. & Th.M. Lexmond (1999). Heavy-metal balances. I: General aspects of cadmium, copper, zinc and lead balance studies in agro-ecosystems. Journal of Industrial Ecology 2 (4): 45-60.
- Richard, T.L. & P.B. Woodbury (1994). What materials should be composted? BioCycle September: 63-68.
- Ross, S.M. (ed.) (1994). Toxic metals in soil-plant systems, John Wiley & Sons, New York, 469 pp.

- Smaling, E. (1993). An agro-ecological framework for integrated nutrient management with special reference to Kenya. Ph.D. dissertation, Wageningen Agricultural University, Wageningen, 250 pp.
- Technische Commissie Bodembescherming (TCB) (1991). Advies kwaliteit en gebruik van GFT-compost. TCB-rapport A90/04. Leidschendam, 32 pp.
- U.S. Environmental Protection Agency (US EPA) (1993). Standards for the use or disposal of sewage sludge. Federal Register 58: 9248-9415.
- Van Riemsdijk, W.H., Th.M. Lexmond, C.G. Enfield & S.E.A.T.M. van der Zee (1987). Phosphorus and heavy metals: Accumulation and consequences. In: H.G. van de Meer et al. (eds.): Animal manure on grassland and fodder crops. Martinus Nijhoff Publishers, Dordrecht, pp. 213-227.
- Van Straalen, N.M. & C.J. Denneman. 1989. Ecotoxicological evaluation of soil quality criteria. Ecotoxicology & Environmental Safety 18: 241-251.

II.5 *Dynabox*: A dynamic multi-media fate model with applications to heavy metals

Reinout Heijungs

II.5.1 Introduction

Many "integrated economic-environmental models" comprise a more or less sophisticated modelling of the economic subsystem, distinguishing various industries and households, describing import and export, and sometimes explicitly dealing with stock forming and time-lag in production and consumption (*e.g.*, Victor, 1972; Perrings, 1986). However, with respect to the environmental subsystem, these models almost always stop at the level of emissions, and do not explicitly address the subsequent pathways that pollutants traverse after they have left the economic subsystem. One reason is probably a misfit between the economy-wide character of most economic models, necessitating the lumping of pollutants with a subsequent loss of details, and the highly regionalised character of most environmental models, requiring very detailed information on nature and release pattern of pollutants. This chapter discusses the use of generic environmental models to make an assessment of potential hazards for men and the environment on the basis of moderately specific data, and can hence be seen as an attempt to bridge the gap.

A break-down of the mechanisms inside the environmental subsystem would require at least the following elements (*cf.* Van Leeuwen & Hermens, 1995):

- *exposure assessment*, *i.e.* the step from emission of pollutants to concentration in the environment or intake by or exposure of target organisms;
- *risk or damage assessment*, *i.e.* the step from concentration in the environment or intake by or exposure of target organisms to damage to the environment or target organisms;
- *value assessment*, *i.e.* the step from damage to the environment or target organisms to a judgement involving ethical considerations (like: birds are more important than worms).

This chapter is primarily devoted to the exposure assessment: it thus describes the *fate* of chemicals in the environment. Some aspects of the damage assessment are dealt with as well. The issue of value assessment is outside the scope of this chapter.

II.5.2 Goal and scope of the model

This section starts with a concise overview of the risk assessment procedure and the idea behind the models that are used to support risk assessment of chemicals. The exposure assessment is the first topic of interest. The damage assessment is a second topic.

A chemical that enters the environment will in general not stay there. It will move to other compartments (*e.g.* from the atmosphere to the soil), it will disappear (*e.g.* by immobilisation in the sediment), and it will enter plants, animals and/or human beings (*e.g.* through drinking contaminated water),. All these of classes of transport phenomena may be described as environmental processes. A more systematic catalogue of environmental processes is the following:

- diffusive transport, through thermodynamic partitioning due to a chemical disequilibrium;
- advective transport, along with physical flows, such as river flows and rainfall;
- degradation, often by chemical reactions, but also by biological processes and photolysis;
- immobilisation, for instance by burial in the deep sediment;
- intake, for instance through respiration, consumption of food and water, and through the skin.

The simultaneous taking-place of these environmental processes determines the fate of a chemical in the environment.

It will be clear that the fate of a chemical is extremely decisive in determining whether or not the chemical will exert a harmful influence on the environment, and, if so, on the magnitude of this influence. For instance, heavy metals have a much larger interaction with the lungs than with the digestive track. It is therefore of enormous interest to know to what extent a released heavy metal will reside in the air. Also, a substance like toluene is quite toxic but it has a short lifespan due to chemical instability on the other hand. There are many substances that are at least as toxic as heavy metals. One of the problems of heavy metals is that they share a relatively high toxic potential with an extremely long chemical stability. It is therefore essential to consider the fate of heavy metals in some detail.

The environmental processes that determine the fate of a chemical may of course be the subject of a modelling exercise. A widely recognised class of suchlike models is the multi-media fate model (see, *e.g.*, Cowan *et al.*, 1995). Multi-media fate models incorporate a number of environmental compartments (air, water, soil, groundwater, *etc.*) and describe the flows of a chemical between these compartments and the degradation within each compartment by means of mathematical equations for each environmental process. Figure II.5.1 illustrates what is involved in a multi-media model.

The structure of these model equations is in general a system of linear differential equations, based on mass-balance conditions and first-order kinetics. Multi-media fate models assume that there is a homogeneous mixing within every compartment, so that there are no local concentration gradients. This is a bold assumption, but it is certainly one that facilitates the modelling step.

As an example, consider a two-compartment model comprising air and water (taken from Heijungs, 1995). It is assumed that there is degradation of the chemical in air, and that it is proportional to the amount that is present in air, the coefficient of proportionality being $kdeg_{\text{air}}$. Further, it is assumed that there is a direct emission of the chemical to air of magnitude $EMIS_{\text{air}}$. And the exchange between air and water is assumed to be proportional with the difference in concentration between the two compartments, the coefficient of proportionality being λ. For the aquatic compartment, a similar reasoning can be held. The system of differential equations that govern the fate in this simple 2-media model is thence:

$$\begin{aligned} \frac{d\overline{m}_{air}(t)}{dt} &= -kdeg_{air}V_{air}C_{air}(t) + EMIS_{air} + \lambda(C_{water}(t) - C_{air}(t)) \\ \frac{d\overline{m}_{water}(t)}{dt} &= -kdeg_{water}V_{water}C_{water}(t) + EMIS_{water} + \lambda(C_{air}(t) - C_{water}(t)) \end{aligned} \tag{12}$$

Figure II.5.1. Overview of a multi-media fate model with the compartments air, water, sediment and three types of soil, with an indication of the different transport and degradation processes.

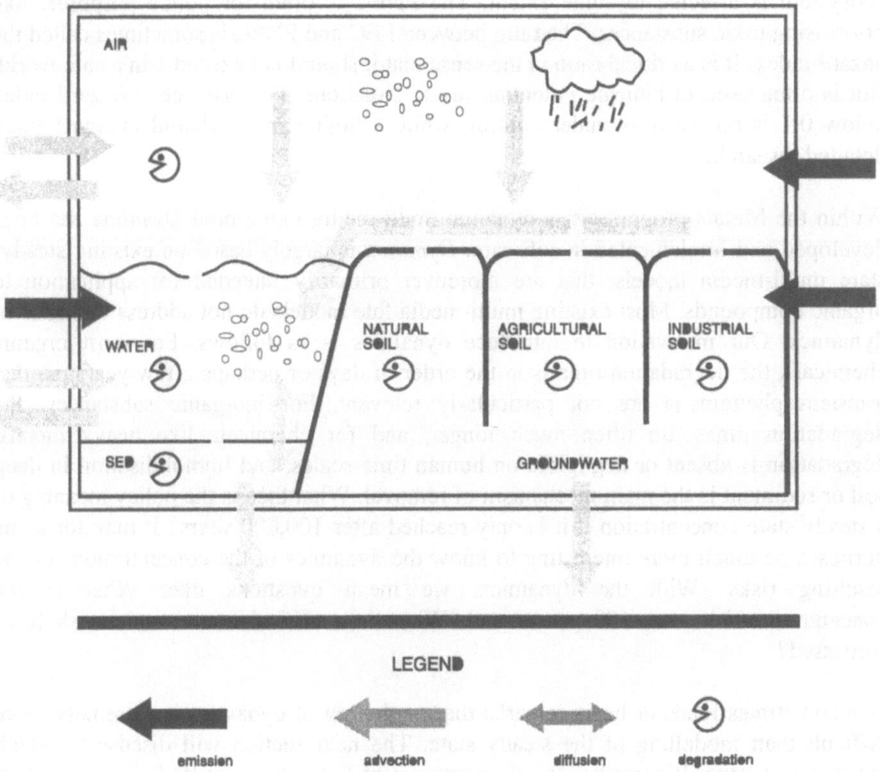

For reasons that we will discuss later in this chapter, multi-media fate models are almost exclusively used to calculate steady-state concentrations. In that case, the left hand sides of the differential equations are put to zero, and the system of equations is solved for C_{air} and C_{water}. This can easily be done by matrix inversion.

A more realistic multi-media model includes several compartments and many environmental processes. The structure, however, is similar to the one outlined in the example. The incorporation of exposure routes from the concentrations in air, water, *etc.* is straightforward when data on respiration, consumption of drinking water and food are known.

Concentrations in environmental compartments and intake by human beings are already much more meaningful than plain emission flows. But still, there is a large difference between 1 mg/m^3 mercury in the air and 1 mg/m^3 copper in the air. To account for these differences in intrinsic toxicity or hazard, risk assessment procedures extend a damage assessment to the exposure assessment.

A popular approach is the so-called PEC/PNEC approach. The predicted environmental concentration (PEC) is the concentration that is the output of the multi-media fate model, so C_{air}, C_{water}, *etc.* The predicted no-effect concentration (PNEC) is the concentration at which it is believed that no more than 5% of the species of an ecosystem is affected to some extent. The PNEC is often for policy purposes, like prioritising toxic substances. The ratio between PEC and PNEC is sometimes called the hazard index. It is a critical ratio in the sense that it should not exceed 1 in a safe world. But is often used to indicate potential unsafe situations. For instance, a hazard index below 0.1 is regarded as under control, while a higher index should instigate more detailed research.

Within the Metals programme, a dynamic multi-media fate model *Dynabox* has been developed and implemented in software. *Dynabox* is largely based on existing steady-state multi-media models, that are moreover primarily intended for application to organic compounds. Most existing multi-media fate models do not address the issue of dynamics. Our motivation to introduce dynamics is as follows. For most organic chemicals, the degradation time is in the order of days or perhaps a few years, so that transient phenomena are not particularly relevant. For inorganic substances, the degradation times are often much longer, and for chemicals like heavy metals, degradation is absent or negligible on human time scales, and immobilisation in deep soil or sediment is the main mechanism of removal. What then is the policy meaning of a steady-state concentration if it is only reached after 100,000 years? It may for some purposes be much more interesting to know the dynamics of the concentrations or the resulting risks. With the dynamics, we mean questions like: What is the concentration/risk after 100 years? and: When is a critical concentration/risk level surpassed?

One sometimes reads or hears remarks that modelling of dynamics is extremely more difficult than modelling of the steady state. The next section will discuss to which extent this is true. It describes the development of a dynamic model from an existing steady-state model. And it will apply the newly developed model to a small number of metals.

II.5.3 Modelling principles and required data

A dynamic multi-media fate model has been developed on the basis of the steady-state model that is integrated in VROM (1994), which on its turn is based on Van de Meent (1991). This latter model, labelled *Simplebox*, also provided the inspiration for the name of the newly developed model: *Dynabox*.[1] Below, the deviations from the original *Uses* model are described: the compartments, choice of parameters for metals, the dynamics, the effect part, and some aspects of implementation. Readers interested in an overview of the basic principles of multi-media fate models are referred to Mackay (1991), Cowan et al. (1995) and to VROM (1994) for more information on *Uses*.

The compartments
The *Uses* model comprises degradation in and transport between a number of compartments: air, surface water, soil, sediment, and so on. The newly developed model is extended to comprise the following compartments:
- the regional model: air, surface water, suspended matter, biota, sediment, natural soil, agricultural sand soil, agricultural peat soil, agricultural clay soil, pore water in sand soil, pore water in peat soil, pore water in clay soil, industrial soil, ground water;
- the continental model: air, surface water, suspended matter, biota, sediment, natural soil, agricultural soil, industrial soil, ground water;
- the sea model: air, seawater, suspended matter, biota, sediment;
- the outside world: deep soil, deep sediment.

The regional model is embedded in a larger system with which it may exchange substances through import and export. The concentrations at the continental and sea system are determined by the user. The outside compartments provide some ultimate sinks with which the regional system is not in steady state, but of which the concentrations are determined by equilibrium partitioning.

Adapting the model for metals
Many parameters that find a place in the multi-media models are based on the intended use for organic degradable chemicals. This creates a barrier in the application of suchlike models for other substances, like metals. There are several reasons for this:
- Several models are able to estimate lacking parameters by means of other parameters or as a default, but only under certain restrictions which are often invalid for metals. In *Uses* 1.0, for instance, the default value for photo-degradation in air is 160 days, and the estimation of bio-concentration factors provide values that are quite a bit off the empirical values.
- Certain models prevent the user for making certain mistakes by refusing to accept values that are improbable for organic substances. For instance, the degradation

---

1 Meanwhile, a newer version of *Simplebox* has appeared (*Simplebox 2.0*: Brandes *et al.*, 1996), as has newer versions of *Uses* (*Uses* 2.0; Linders & Jager, 1997; *Euses 1.0*: EBC, 1997). These developments, however, came too late to be incorporated into *Dynabox*. It is doubtful whether it would make much difference for the case of heavy metals, because *Euses 1.0* is a refinement of *Uses 1.0* that does not address the case of heavy metals and *Uses 2.0* is mainly an improvement for pesticides.

times should be at most 10,000 days (= 27 years) in *Uses* 1.0, which is quite unrealistic for metals.

- The calculation often involves parameters that are ill-defined for metals. For instance, the vapour pressure is essential in *Uses* 1.0, while it is undefined for metals.

The first point may be circumvented by putting more effort in data collection, so that estimation of missing parameters is not necessary. The second point could be solved by not implementing the restrictions on the domain of parameter values in *Dynabox*. The third point, however, is more difficult to avoid. One can try the effect of putting parameters to, say, 10^{20} and checking if a value of 10^{30} makes no significant difference. A useful reference in this context is Crommentuijn et al. (1997).

From steady-state model to dynamic model

In general, multi-media-fate models calculate the steady-state concentrations of a pollutant in different compartments. As mentioned before, a policy-relevant application to metals requires a dynamic calculation. This subsection describes how such a dynamic model is obtained from a steady-state model.

a. The steady-state model

We first describe the essentials of the steady-state model. This is done quite extensively, because the connection to a dynamic model would otherwise be unclear, and because most texts on multi-media models are less explicit in the mathematical structure of the general set-up.

The heart of the multi-media fate model is a differential equation that expresses a mass balance condition for each considered medium (compartment), i.e. air, water, soil, ground water, sea, etc.):

$$\frac{\mathrm{d}m_i(t)}{\mathrm{d}t} = \left(\frac{\mathrm{d}m_i(t)}{\mathrm{d}t}\right)_{\text{from other media}} - \left(\frac{\mathrm{d}m_i(t)}{\mathrm{d}t}\right)_{\text{to other media}} + \left(\frac{\mathrm{d}m_i(t)}{\mathrm{d}t}\right)_{\text{through emission and import}} - \left(\frac{\mathrm{d}m_i(t)}{\mathrm{d}t}\right)_{\text{through degration and export}}$$

where $m_i(t)$ denotes the mass of a the substance under study in compartment i at time t. Three of the four terms on the right hand side of the equality-sign are constructed according to the principles of first-order kinetics:

$$\left(\frac{\mathrm{d}m_i(t)}{\mathrm{d}t}\right)_{\text{from other media}} = \sum_j \left((ADV_{ij} + DIFF_{ij}) \times C_j(t)\right) \qquad (13)$$

$$\left(\frac{\mathrm{d}m_i(t)}{\mathrm{d}t}\right)_{\text{to other media}} = \left(\sum_j (ADV_{ji} + DIFF_{ji})\right) \times C_i(t) \quad (14)$$

$$\left(\frac{\mathrm{d}m_i(t)}{\mathrm{d}t}\right)_{\text{through degradation and export}} = \left(V_i \times kdeg_i + (ADV_{\text{outside},i} + DIFF_{\text{outside},i})\right) \times C_i(t) \quad (15)$$

while the fourth one contains two direct source terms:

$$\left(\frac{\mathrm{d}m_i(t)}{\mathrm{d}t}\right)_{\text{through emission and import}} = EMIS_i(t) + IMPORT_i(t)$$

In these equations, ADV_{ji} denotes a coefficient for advective transport from compartment j to compartment i, $DIFF_{ji}$ a similar one for diffusive transport, $kdeg_i$ a degradation coefficient for compartment i, V_i the volume of compartment i, $EMIS_i(t)$ the direct emission flow into compartment i at time t, $IMPORT_i(t)$ the transboundary flow into compartment i at time t, and $C_i(t)$ the concentration in compartment i at time t. Furthermore, concentration and mass are related by the volume V_i of the compartment by definition: $C_i = \frac{m_i}{V_i}$. Observe that the coefficients for advective and diffusive transport and for degradation are time-independent (hence the qualification "quasi-dynamic" in Brandes et al., 1996), and that there are besides the masses (or equivalently: concentrations) other quantities that may be varying with time: the emission flow and the import flow due to concentrations in the outside world.

The general model structure is hence of the form

$$\frac{\mathrm{d}\mathbf{m}(t)}{\mathrm{d}t} = \mathbf{L} \bullet \mathbf{m}(t) + \mathbf{f}(t) \quad (16)$$

where $\mathbf{m}(t)$ is a column-vector of all $m_i(t)$ and $\mathbf{f}(t)$ a column-vector of emissions and imports:

$$\mathbf{m}(t) = \begin{pmatrix} m_1(t) \\ \cdots \\ m_i(t) \\ \cdots \end{pmatrix}; \ \mathbf{f}(t) = \begin{pmatrix} EMIS_1(t) + IMPORT_1(t) \\ \cdots \\ EMIS_i(t) + IMPORT_i(t) \\ \cdots \end{pmatrix} \quad (17)$$

Furthermore, $\mathbf{L}$ is a vector of coefficients that determine the fate of the chemical (hence the term fate matrix in Heijungs (1997)):

$$\mathbf{L} = \begin{pmatrix} l_{11} & \cdots & l_{1j} & \cdots \\ \cdots & \cdots & \cdots & \cdots \\ l_{i1} & \cdots & l_{ij} & \cdots \\ \cdots & \cdots & \cdots & \cdots \end{pmatrix} \qquad (18)$$

while for an arbitrary element of **L** we have

$$l_{ij} = \begin{cases} \dfrac{ADV_{ij} + DIFF_{ij}}{V_i} & (i \neq j) \\ \dfrac{ADV_{ij} + DIFF_{ij}}{V_j} - \sum_j \dfrac{ADV_{ji} + DIFF_{ji}}{V_j} - kdeg_i + \dfrac{ADV_{\text{outside},i} + DIFF_{\text{outside},i}}{V_i} & (i = j) \end{cases} \qquad (19)$$

The differential equation (16) is in most multi-media fate models taken as a starting point, but with the addition that a steady-state situation is assumed:

$$\frac{\mathrm{d}\mathbf{m}(t)}{\mathrm{d}t} = \mathbf{0} \qquad (20)$$

so that the time index t then refers to the situation after an infinitely long transition time, while the emission flow is taken to be time-independent:

$$\mathbf{f}(t) = \mathbf{f} \qquad (21)$$

The result is a simple matrix equation:

$$\mathbf{0} = \mathbf{L} \bullet \mathbf{m}(\infty) + \mathbf{f} \qquad (22)$$

which is easily solved by matrix inversion:

$$\mathbf{m}(\infty) = -\mathbf{L}^{-1} \bullet \mathbf{f} \qquad (23)$$

This procedure corresponds to those of *Simplebox* 1.0 (van de Meent, 1991 (p.50-51)), *Simplebox* 2.0 (Brandes et al., 1996 (p.118-119), and, more implicitly, of Mackay (1991, p.177) and *Uses* 1.0 (VROM, 1994 (p.160)). It should be stressed that (23) only makes sense when **L** is invertible. Although this condition is not discussed in the above references, it is not a trivial one; see the appendix of Heijungs (1995).

b. The dynamic model

If we wish to deviate from this normal practice of describing a steady state and seek to describe a dynamic concentration pattern, we need to solve the differential equation (16), either numerically or analytically (cf. Mackay, 1991 (p.183-184), Van de Meent, 1991 (p.57-58) and Brandes et al., 1996 (p.123-125)).

Let us first consider the numerical approach. Simple stepwise integration (Euler's method[2]) yields

---

2 Of course, we may also deal with the topic in a more sophisticated way. Euler's method of integration may be replaced by more accurate methods, like those of Runge-Kutta, and with a

$$\mathbf{m}(t+\Delta t) \approx \mathbf{m}(t)+\frac{\mathrm{d}\mathbf{m}(t)}{\mathrm{d}t}\times\Delta t \qquad (24)$$

By taking Δt sufficiently small, we will be able to approximate the time series $m(t)$, $m(t+\Delta t)$, $m(t+2\Delta t)$, $m(t+3\Delta t)$, ... with a reasonable accuracy. The operational formula is then simply

$$\mathbf{m}(t+\Delta t) \approx \mathbf{m}(t)+\left(\mathbf{L}\bullet\mathbf{m}(t)+\mathbf{f}(t)\right)\times\Delta t \qquad (25)$$

By comparison with the steady-state calculation (equation (23)), we see that there is, besides the more or less arbitrary integration step size Δt, one additional data requirement: $m(t)$, i.e. the amounts (or concentrations) of pollutant in the starting year. Environmental agencies or statistical bureaux often possess these background data for some initial year, usually a few years earlier, like 1990 or 1995.

Another important direction is that of leaving the field of numerical approximation, and trying to find analytical solutions of the differential equation (16). It is not difficult to derive a general solution in which Δt may assume any value, so also non-infinitesimal small values:

$$\mathbf{m}(t+\Delta t) = e^{(t+\Delta t)\mathbf{L}}\bullet\left(\mathbf{m}(t)+\int_{t}^{t+\Delta t} e^{-s\mathbf{L}}\bullet\mathbf{f}(s)\mathrm{d}s\right) \qquad (26)$$

This expression, however, is difficult to work with in practice for two reasons: it contains an integral that needs to be computed for the specific emission pattern $\mathbf{f}(t)$, and it contains an exponent with a matrix. If we simplify the problem to the extent that a constant emission flow is assumed, we are able to solve the integral and obtain

$$\mathbf{m}(t+\Delta t) = e^{(t+\Delta t)\mathbf{L}}\bullet\left(\mathbf{m}(t)+\mathbf{L}^{-1}\bullet\mathbf{f})\right)-\mathbf{L}^{-1}\bullet\mathbf{f} \qquad (27)$$

but are still left with a matrix in the exponent. Although possibilities exist to deal with this (see, e.g., Apostol, 1969 (p.201 ff.) and Brandes et al., 1996 (p.124-125)), the mathematics indeed gets rather involved (cf. Mackay, 1991 (p.184)), and it will do so even more when the assumption of constant emission flows is dropped. The current version of *Dynabox* is based on Equation (13).

Summarising the differences between steady-state and dynamic models for different classes of people:

- the *user* needs to specify one additional data item: the amounts or concentrations of pollutant in the different environmental compartments (air, water, etc.);
- the *user* will be confronted with a problem of interpretation: instead of one result, a whole time series of results will be obtained;
- the *modeller* that constructs the multi-media fate model sees no difference: it is his or her task to specify the matrix **L**, and that matrix is the same for both types of model;

---

possible adaptive stepsize Δt (see, *e.g.*, Press *et al.*, 1982). This is, however, beyond the scope of the present chapter.

- the *software engineer* needs to design appropriate routines for numerically integrating systems of differential equation, or to go into the theory of exponents with matrices.

The effect part
As stated in the Introduction, this chapter and *Dynabox* primarily describe the exposure assessment. However, in order to give an indication of damage, an effect module (cf. Jager & Visser, 1994) has been added. It is based on the "PEC/PNEC" approach, where "PEC" stands for the predicted environmental concentration, and "PNEC" denotes the predicted no-effect concentration. For humans, the ratio of the predicted daily intake and the acceptable daily intake (PDI/ADI) is used. These ratios are also referred to as the hazard quotient (and their reciprocal value as the margin of safety). In theory, a hazard quotient below 1 is an indicator for a safe situation, while a value larger than 1 indicates a hazardous situation. The real interpretation is obviously different, for instance, since a value below 1 does not necessarily indicate a safe situation due to uncertainties in substance parameters, environmental parameters, model assumption, and so on. The hazard quotient must therefore be regarded as an indicator that is the outcome of a screening tool.

Despite these words of caution, the results in Parts III and IV are simply based on the PEC/PNEC, without sensitivity analysis, and without further study in a more detailed risk analysis. As a justification for this we might say that the entire programme was of an explorative nature, and that all results provide mere indicative trends and no hard facts. The model calculates PEC/PNEC values for aquatic and terrestrial ecosystems, as well as PDI/ADI values for human beings. Different types of PNEC for ecosystems can be introduced, such as the reference value, the limit value and the maximum permissible concentration.

Implementation in software
The model has been implemented in a software code. The source code was written in Turbo Pascal. An executable code runs on MS-DOS, and is freely available.[3] For reasons of comparison with the original *Uses* 1.0, an option has been incorporated, which allows the user to investigate the differences in modelling results between the original and modified multimedia fate model. A further feature that deserves to be mentioned is an option that enables the user to choose between steady-state computation (Mackay's level III) and a dynamic computation (level IV).

II.5.4 Results and interpretation

Results of the model for real-world situations on cadmium, copper, lead and zinc can be found in Parts III and IV and in Guinée et al. (in press). Comparison with the *Uses* model yields a large difference for the steady-state values, mainly due to *Uses*' limitation of the degradation time to 27 years. With this value, *Dynabox* computes

3 Contact the web-site http://www.leidenuniv.nl/interfac/cml/ssp/publssp.html

results that are quite similar to that of *Uses*. The differentiation of agricultural soil into sand, peat and clay does not affect the overall picture. However, it introduces differences in soil concentrations of a factor up to 10. The results for soil are in reasonable agreement with those of Moolenaar (1998). A more complete discussion of the results of *Dynabox* is in preparation (Heijungs et al., in prep.).

II.5.5 Links to other models

Dynabox needs emission flows as an input. The emission data are entered through a menu. Emission data may, however, also be read from an external file, for instance, produced by a model of the economic subsystem. A link with Flux (see Section II.2) is thus easy to realise. But also connections with the national statistics (input-output tables, emission inventories, etc.) are in principle possible.

The model gives concentrations in environmental media (air, surface water, etc.) and intake of humans as an output, or PEC/PNEC-like hazard quotients. Concentrations/intakes may be calculated as a steady-state value, at any moment in time, or the whole time series may be exported. Any of these forms of output may become the input of valuation models. This aspect has, however, not been studied.

References

- Apostol, T.M. (1969). Calculus. Volume II. Multi-variable calculus and linear algebra, with applications to differential equations and probability. Second edition. John Wiley & Sons, New York.
- Brandes, L.J., H. den Hollander & D. van de Meent (1996). SimpleBox 2.0. A nested multimedia fate model for evaluating the environmental fate of chemicals. RIVM, Bilthoven.
- Cowan, C.E., D. Mackay, T.C.J. Feytel, D. van de Meent, A. Di Guardo, J. Davies & N. Mackay (1995). The multi-media fate model. A vital tool for predicting the fate of chemicals. SETAC, Pensacola.
- Crommentuijn, T. M.D. Polder & E.J. van der Plassche (1997). Maximum permissible concentrations and negligible concentrations for metals, taking background concentrations into account. RIVM, Bilthoven.
- Guinée, J.B., J.C.J.M. van den Bergh, J. Boelens, P.J. Fraanje, G. Huppes, P.P.A.A.H. Kandelaars, Th.M. Lexmond, S.W. Moolenaar, A.A. Olsthoorn, H.A. Udo de Haes, E. Verkuijlen & E. van der Voet (1999). Evaluation of risks of metal flows and accumulation in economy and environment. Ecological Economics, in press.
- Heijungs, R. (1995). Harmonization of methods for impact assessment. Environmental Science & Pollution Research 2:4, pp 217-224.
- Heijungs, R. (1997). Economic drama and the environmental stage. Formal derivation of algorithmic tools for environmental analysis and decision-support from a unified epistemological principle. PhD thesis Leiden University.
- Heijungs, R, A. Wegener Sleeswijk & J.B. Guinée (in prep.). Dynabox. A generic dynamic multi-media fate model.

- Jager, D.T. & C.J.M. Visser (1994). Uniform system for the evaluation of substances (USES), version 1.0. VROM, 's-Gravenhage.
- Leeuwen, C.J. van & J.L.M. Hermens (Eds.) (1995). Risk assessment of chemicals. An introduction. Kluwer Academic Publishers, Dordrecht.
- Linders, J.B.H.J. & D.T. Jager (Eds.) (1997). Uses 2.0. The uniform system for the evaluation of substances, version 2.0. RIVM, Bilthoven.
- Mackay, D. (1991). Multimedia environmental models. The fugacity approach. Lewis Publishers, Chelsea.
- Meent, D. van de (1993). Simplebox. A generic multimedia fate evaluation model. RIVM, Bilthoven, the Netherlands.
- Moolenaar, S.W. (1998). Sustainable management of heavy metals in agro-ecosystems. PhD thesis Landbouwuniversiteit Wageningen.
- Perrings, C. (1987). Economy and environment. A theoretical essay on the interdependence of economic and environmental systems. Cambridge University Press, Cambridge.
- Press, W.H., B.P. Flannery, S.A. Teukolsky & W.T. Vetterling (1989, 1992). Numerical recipes in Pascal. The art of scientific computing. Cambridge University Press, Cambridge.
- Victor, P.A. (1972). Pollution. Economy and environment. George Allen & Unwin Ltd., London.

II.6 Sustainability indicators for the case of heavy metals

Ester van der Voet, Jeroen B. Guinée & Helias A. Udo de Haes

II.6.1 Introduction

Materials and Substance Flow Analysis (MFA/SFA) studies are designed to support environmental decision-making. Although in practice such studies have been carried out successfully (e.g. Adriaanse et al., 1997; Kleijn et al., 1997), the issue of connecting such research with policy has arisen frequently (Brunner et al., 1997). Researchers may feel the results of MFA and SFA studies are giving a clear message, but for policy-makers they are often not quite so self-evident. It would therefore seem appropriate to pay closer attention to the translation of MFA/SFA results into policy-relevant terms. Three issues need to be addressed for such communication: the basic principles of MFA and SFA, the terminology and the complexity of MFA/SFA results.

Regarding the basic principles, the usefulness of investigating societal metabolism is sometimes questioned. Policy-makers often feel that knowledge of emissions and extractions is sufficient. Many publications have been devoted to the importance of studying societal metabolism since the publication of the concept of industrial metabolism by Ayres (1989) and the political awareness of the role of societal flows and stocks as the instigators of environmental problems is slowly growing in the elaboration of policy principles such as 'integrated chain management' (Anonymous, 1991).

Terminology is always a difficult issue in a relatively new area of investigation such as MFA. Even among scientists there is no established terminology in this area, which often leads to confusion. On top of that there is the lack of coherence between the scientific and the policy vocabulary. We can see, for example, that the policy concept of 'sustainability' has found its way into MFA research, but that it has become such a very broad concept that it covers virtually everything and has therefore been stripped of any real meaning. In order to close the gap between policy and science, a more specific connection must be made: MFA scientists should point out more clearly the relevance of their results in terms of policy means and ends.

The third obstacle is the complexity of the MFA results. These results are, in most cases, comprised in an overview of flows and/or stocks connected with a given region. Often, such an overview is too complicated to distil precisely the relevant information. A further interpretation of the overview data is then required, also to avoid the risk for deriving spurious conclusions. In the MFA framework as presented by Udo de Haes et al. (1997), this 'interpretation' is the third element after 'goal and systems definition' and 'inventory and modelling'. The present article is dedicated to this third element of the MFA framework, which has received relatively little attention so far in a methodological sense. A choice has been made to elaborate this by defining a *set of indicators*. By doing so we hope to achieve different aims at the same time: reducing complexity and establishing a better connection with the language of policy. In this direction, the 'socio-ecological' indicators of Azar et al. (1996) can be mentioned. A further specification is made by zooming in on SFA, and especially on the case of heavy metals.

Indicators play an important role in the interpretation of environmental data for environmental policy. The general idea is to aggregate the rather large and ungainly dataset into a limited number of measures or yardsticks relevant for environmental policy. Indicators are widely used by policy-makers to measure developments in the state of the environment, human influence on the environment and the effectiveness of chosen policy measures. The concept 'indicator' is not strictly defined and in practice many widely differing things may serve as indicators. Several attempts have been made to establish a classification of indicators (e.g. Opschoor & Reijnders, 1991; OECD, 1993a and 1993b).

The indicators developed during this research programme are designed to provide information with regard to the flows and stocks of heavy metals of relevance for an integrated substance chain management policy. Substance chain indicators must be constructed in such a way as to provide information regarding:

- the existence and causes of environmental problems related to the substance
- the management of the substance chain or cycle in society
- early recognition of future problems
- the influence of policy measures, including both their effectiveness and various types of problem-shifting.

These demands imply the need for the definition of *reference values* indicating a desired or sustainable level for the individual indicators. For each indicator the availability of such reference values and the approach taken if reference values are missing are discussed below.

In addition, requirements can be defined for the indicators as a group, which must be suitable for evaluating an SFA overview for a specific year but also for evaluating changes in flows and stocks over time as well as alterations thereof, as induced by environmental policy, for example. Therefore, a comparison between different regimes must also be possible. Rather than designing a weighting procedure for the indicators in order to arrive at one score for each scenario, in Section IV we have opted for qualitative evaluation, using the sum total of information provided by the available indicators.

We have defined indicators for environmental flows and stocks and indicators for flows and stocks in the economy. Indicators in the first group are related to the core issue of most substance-oriented policies: *environmental quality* and human influence on it. Indicators in the second group relate to a society's metabolism and can therefore be regarded as indicators for *(sustainable) development*. In the context of the environment they may fulfil an *early-warning* function. Some indicators of this group refer to the broader impacts of a society's metabolism and may therefore have a function in detecting *problem-shifting*.

Below, the indicators applied to the case of heavy metals are explained briefly; for a more extensive description we refer to Van der Voet et al. (1998). For several of the indicators, processes in the economy need to be allocated to a life cycle stage (extraction and refining, production, use and waste treatment) and flows and stocks need to be characterised into a number of categories. The definition of the various stock and flow categories and a mathematical definition of the indicators are presented in Van der Voet et al. (1997).

The indicators are grouped according to the research questions stated in Section I.1:

- research question 3: indicators for the fate of the mined metals
- research question 4: indicators for evaluation of present management in terms of sustainability
- research question 5: indicators for design of a sustainable management regime.

II.6.2 Indicators for the fate of the mined metals

Indicators for the fate of the mined metals relate to the losses from the economic system. These may be undesirable losses, such as emissions to the environment, but also acceptable forms of loss, such as immobilisation. In systems out of equilibrium accumulation may take place, which is not a loss but may account for differences between inflow and outflow. The indicators developed for the fate of the metals are:

1. total emissions
2. total landfill
3. accumulation in the economy
4. pollution export

Obviously, an indicator for immobilisation or other acceptable forms of loss is lacking. Defining what is acceptable proved to be the main stumbling block. At present, no form of waste disposal is considered acceptable. Futuristic solutions such as vitrification or refilling the mines might apply for this status.

1. Total emissions (mass/year)

The indicator is the *total amount of emissions within the region*. Below, several indicators are defined regarding the *relative* amount of leakages or cycle losses, which is relevant information from the point of view of formulating policy measures. However, it is the *absolute* amount that is indicative for the environmental problems created by a certain chain management regime. Here, then, the total amount of emissions may serve as an indicator.

<u>Significance of the indicator</u>: What the significance of this indicator is depends on what exactly counts as an emission. In a strict sense, any transgression of the economy/environment border is an emission. However, a narrower definition is generally adopted: 'emissions' are the emissions into the atmosphere and surface waters as well as the diffusive emissions into soils. When considering environmental quality the latter definition may be preferred, since this excludes the landfill sites where large amounts are usually stored but relatively little disperses from the site.

<u>Method of calculation</u>: The emissions can be extracted directly from the overview of flows and stocks.

<u>Interpretation for environmental policy</u>: For the purpose of interpretation emission targets may be used, if available. It may be necessary to break down total emissions into categories, i.e. emissions to the various environmental compartments. Often, emission targets do not exist. In that case, 'less is better' may still be used to compare alternatives. It can be argued that using emission targets is merely a less sophisticated approach to the

environmental problems involved. On the other hand, emission targets often include other policy objectives apart from environmental concern, for example political considerations.

2. Total landfill (mass/year)
The indicator is the *total amount of the metal entering landfill* (mass/year).
Significance of the indicator: The indicator is one of the *fate* indicators, providing information on the ultimate destination of the metals after their period of use in the economy. It is not directly linked to environmental risks, although it may indicate a potential future source of emissions through dispersion from landfill sites.
Method of calculation: The amount of landfill can be extracted directly from the overview of flows and stocks.
Interpretation for environmental policy: No target value is available, but in view of waste prevention as a general target alternative management scenarios can be compared with the adagio 'less is better'. Some forms of landfill might be considered as immobilisation, for example landfill on isolated, controlled sites. Landfill is usually considered preferable to emissions. In the long run, however, emissions from landfill sites may increase and landfills might become important emission sources. Whether this actually occurs and, if so, at what moment in time also depends on aspects of quality regarding the construction and management of such a site. In the best case, when leaching from the site is indeed extremely limited, such landfill may be classified as 'immobilisation'. At present no landfilled metals are considered to be immobilised. This may change in the future with development of immobilisation techniques or use of truly isolated storage sites, located in the geosphere, for example.

3. Accumulation in the economy (mass/year)
The indicator is the *increase in the economic stocks in the region over the year.* Accumulation of substances in the economy takes place in materials and products; accumulation in landfills is not taken into account in this indicator. By far the largest economic stocks can be found in the use or consumption stage of the substance's life-cycle: in products used in households, in materials tied up in buildings, roads, equipment, vehicles, etc.
Significance of the indicator: In itself, accumulation in materials and products does not constitute a problem: there is no leakage to the environment. What accumulation does mean, though, is a lack of equilibrium in the economic system. Growing stocks indicate a risk of leakages and/or waste volumes growing in magnitude in the future. Accumulation in the economy thus functions as a warning signal for future environmental problems.
Method of calculation: This indicator can be extracted directly from the overview of flows and stocks. Specific stocks may be chosen, but the indicator can also be defined on an aggregate level. This latter option may be preferred since it is difficult to indicate 'key' stocks within the economic system.
Interpretation for environmental policy: In contrast to environmental accumulation, economic accumulation cannot be translated directly into policy-relevant terms with the aid of environmental quality standards. It can be stated that, in otherwise comparable systems, more accumulation means a less stable situation (i.e. deviating more from the steady state) and therefore a greater risk of higher emissions in the future. The indicative value is thus relative, not absolute. The absolute value of this accumulation is difficult to interpret. A comparison with other flows or stocks may yield more meaning and enhance comparability between the various metals. Possibilities here are: the total

inflow into the economy (% of inflow accumulating) and the present size of the stock (% stock increase).

4. Pollution export (dimensionless)
The pollution export indicator is *the shifting of pollution problems away from the region* in relation to developments within the region. Several studies have analysed the process of the 'cleaning up' of regions during the last decades (Ayres & Rod, 1986; Stigliani & Anderberg, 1992). One of the possible side-effects of such a process is to shift environmental problems to other areas. Especially in regions with rather strict environmental regulations, the more polluting stages of the life cycle may be relocated outside the area, thus shifting the environmental problems elsewhere. At the global level this may even lead to more pollution.
Significance of the indicator: The pollution export is a measure of the occurrence of benefits and problems associated with use of a substance within the same area, or, in other words, the (implicit) problem-shifting practices of a region.
Method of calculation: In specifying a region's substance life cycle there are two possible points of departure, each resulting in a different idea of what does and what does not belong to the cycle. The two approaches can be characterised as 'regional' and 'functional' (see also Van der Voet, 1995a). It is precisely the difference between those definitions that is relevant for the pollution export indicator.
Taking the regional approach, a picture emerges of the pollution occurring within the region, identical to indicator 1 (total emissions). The functional approach starts from consumption in a given region and specifies the emissions from all processes - up-chain, i.e. extraction and production, as well as down-chain, i.e. waste management - connected with that consumption, whether they occur within or outside the region. This approach bears a certain resemblance to what Wackernagel & Rees (1996) call the *ecological footprint*. They refer to the fact that an urban region requires much more space than its actual territory in order to provide for its inhabitants. In the case of the *pollution footprint*, 'space' must be taken metaphorically.
Interpretation for environmental policy: A reference value of 0% pollution export, indicating that the region's pollution footprint matches the pollution occurring within the region, would seem a logical choice. However, the problem-exporting or problem-importing practices of a region cannot be judged absolutely. Another possibility is the 'less-is-better' approach, not applied to the pollution export but to the pollution footprint: the smaller a region's pollution footprint the better.

II.6.3 Indicators for evaluation of present management in terms of sustainability

Indicators to evaluate the present metals management regime in terms of sustainability can be defined in two directions: pollution problems and depletion problems. By our definition in Section I.2, pollution problems have been translated into terms of human and ecosystem health risks. Therefore, a certain management regime is considered unsustainable if accepted standards for health risks are transgressed. The depletion problem is not considered in this study. Nevertheless, for reasons of completeness, we have defined an indicator measuring the depletion of metals resources. This indicator has not been used in the calculations of Part III and IV. The following indicators are defined:

5. environmental concentration *(PEC/PNEC)*
6. human intake *(PDI/TDI)*
7. environmental accumulation
8. depletion rate.

5. Environmental concentration (mg/kg)

The indicator is the *concentration of a substance in an environmental compartment.* It is a translation of the substance's stock. In addition, a concentration *rise* is a translation of the stock's increase or accumulation (see Indicator 7).

Significance of the indicator: The calculated concentration of a substance in an environmental compartment is a measure for the (potential) loss of environmental quality. It may indicate a human health risk through specific exposure routes, it may indicate a loss of economic functions, for example agriculture or recreation, and it may also indicate a deterioration of ecosystems. For the several functions or values, separate reference values may be defined.

Method of calculation: Through the multimedia model DYNABOX (see Section II.5), the emission outcome of Flux (see Section II.2) is translated into predicted concentrations in environmental media. In addition to the overview of flows and stocks, information is required regarding the 'amount of environmental compartment' in the region under study. This information is included in Dynabox for the Dutch situation.

Interpretation for environmental policy: The introduction of reference values in the shape of concentration standards opens opportunities to evaluate flows in an absolute sense. In the case of heavy metals, we use *PNEC* (Predicted No-Effect Concentration) values as a reference. The *PEC/PNEC* (Predicted Environmental Concentration divided by Predicted No-Effect Concentration) indicates the potential risk (values > 1) of emissions to ecosystem health; this will be referred to further as the *PEC/PNEC* risk ratio. The predicted concentrations are then compared with reference values developed for the Dutch situation and generally used in Dutch environmental policy (Crommentuijn et al., 1997). These reference values have been developed specifically for ecosystem health. In the case study on agriculture *PEC/PNEC*-like indicators have been developed based on a much more refined, single-medium soil model. For more details we here refer to Moolenaar et al. (1997a).

6. Daily intake (mg/day)

The indicator is the *average daily human intake of the substance by all ingestion routes.*

Significance of the indicator: The daily intake may serve as an indicator for the human health risk.

Method of calculation: Using risk-assessment methods, environmental concentrations can be translated into an average daily intake of the substance for humans (for example, Paustenbach 1989). Such methods calculate intake by all the various exposure routes, so that all environmental compartments are included. For other species, daily intake may be calculated in a similar manner. The *PDI* is also obtained from DYNABOX, which includes a risk assessment model, by calculating human intake via the various environmental routes (inhalation, water, food and others) starting from the aforementioned environmental concentrations.

Interpretation for environmental policy: The human daily intake can be compared with ADI or TDI values, such as defined by the WHO. These values refer to a maximum acceptable intake level, and any transgression of these indicates a health risk. A *PDI/TDI* (Predicted Daily Intake divided by Tolerable Daily Intake) with a value > 1 indicates that the metals management regime in question implies a risk to human health. The *TDI* values originate from the WHO when available, or are values developed by RIVM and generally used in Dutch environmental policy.

7. Environmental accumulation (mass/year)
Environmental accumulation is *the increase of a certain environmental stock over the year.* It can be defined either as an aggregate (total environmental accumulation) or on a detailed (accumulation in one specific stock or compartment) level.
The significance of the indicator: Environmental sinks, i.e. stocks in which substances tend to accumulate, are soil, sediment and groundwater. Accumulation seldom indicates a specific environmental problem. What it does indicate, however, is the fact that under the current substance management regime the environment is off-balance. If this regime is continued over the years, the environmental concentration will rise, eventually transgressing quality standards. Moreover, the outflows from such a growing stock often tend to increase and lead to problems in other environmental compartments. In the end, the situation will stabilise at a new (possibly undesirable) steady state.
Method of calculation: This indicator can be extracted directly from the overview of flows and stocks.
Interpretation for environmental policy: In general, it can be stated that the larger the accumulation, the severer will be the state of imbalance and the higher the risk of future problems. As a reference value, zero accumulation may be adopted, in line with the motto 'less is better'. In some cases, a negative accumulation (stock decrease) might even be desirable. However, we may find ourselves entangled in the problem of chemical time bombs (Stigliani & Salomons, 1993): a stock decrease may imply increasing availability of substances formerly locked safely away in stocks. To enhance comparability, environmental accumulation can be related to the total environmental inflow, comparable to the treatment of accumulation in the economy (see Indicator 2). Another possibility is to calculate the point in time that, through ongoing increase in stocks and therefore in concentrations, the *PNEC* will be violated; this is the *PEC/PNEC transition period.* It indicates the urgency of the problem: the shorter the period to transgression, the more pressing is the need to change the management regime.

8. Depletion rate
The indicator is *the amount of extraction of a substance from the environment on behalf of a region, compared with the global extraction of that substance.*
Significance of the indicator: The indicator specifies the *contribution of the region to global resource depletion.* It provides information on the depletion of global resources resulting from the substance management regime within the region.
Method of calculation: Virgin material requirements are determined by the region's consumption, i.e. the use flow (see indicator 7) and by the amount of recycling (see indicator 8). On that side, the information is already available from the other indicators. This information must be combined with information on global extraction of the resource to bring it into perspective. Even more perspective can be obtained by calculating this indicator on a per capita basis and then comparing it with the global per capita average.

Interpretation for environmental policy: For this indicator, it is very difficult to define a reference value. Once again the 'less is better' criterion can be used. For the per capita variant, the global average may be used as a reference value.

II.6.4 Indicators for design of a sustainable management regime

Indicators for the design of a sustainable management regime are concerned with the flows in the economic subsystem. They have no direct relevance for risks or environmental flows, or even for the sustainability of present management. Instead they may be used in the process of designing a substance flow management regime. For example, if the environmental indicators give the message that emissions must be reduced, the indicators from this group may offer options on how to do so: which flows or stocks should be primarily adapted, can an emission reduction be achieved by technical means or by raising efficiency, or should we look at substitution for a possible solution. A number of indicators are defined in this category. However, not all of these could be operationalised sufficiently to apply to the case of the four metals. In the end, the only indicators we succeeded in applying are the first two, below. This implies that the aspect of design of a sustainable management regime is treated less formally in Section IV than evaluation of present management in terms of sustainability. The proposed indicators are:

9. technical efficiency
10. recycling rate
11. use level
12. economic dissipation
13. disturbance rate.

9. Technical efficiency (%)

The indicator is the *technological efficiency of the economic processes within the region.* The efficiency is in fact the converse of the leakages: it is the *fraction of the inflow that ends up in another economic process.*

Significance of the indicator: The efficiency of a comprehensive [** of: 'coherent' = samenhangend?] group of processes indicates the appropriateness of the processes and techniques involved. The factors determining this appropriateness vary with the life cycle stage. For production processes, the indicator points to the possibilities for closing the cycle by technical means: the adoption of more efficient production technologies, or better application of those currently employed. For product usage, technical efficiency is related to two aspects: the life span of the applications, and the percentage of the household waste collected to be treated further in one way or another. For waste management, efficiency is determined by the amount of the substance recovered as secondary materials, and by the amount degraded or immobilised.

Method of calculation: Process efficiency is mostly calculated as OUT/IN, whereby the losses to the environment are extracted from OUT. [** hier, zoals elders, een verwijzing naar 'Section ..', of is dit gewoon 'algemeen'?] A choice has been made to establish efficiency for four subsystems, representing life cycle stages: (1) extraction and refinery, (2) production and manufacturing, (3) use and consumption and (4) waste management. For each subsystem an efficiency percentage can be calculated. (For a justification of this choice and for a further elaboration on the method of calculation and the interpretation, see Van der Voet et al., 1997.)

Interpretation for environmental policy: As a general rule it can be stated that - quite apart from economic considerations - the higher the efficiency of the processes the better. An efficiency of 100%, although this can never be realised, may therefore serve as a reference value. Furthermore, the life-cycle stage responsible for the greatest losses can be identified. In various case studies, it has been shown that the largest leakages no longer occur in the production stage, but in the use or waste management stage, depending on the substance and the application. Also the efficiencies of different sectors within the economy or of the chains of different substances can be compared, as shown in Figure 1 for six heavy metals in the Netherlands.

10. Recycling rate (%)

The recycling rate within the region refers to the *relative amount of generated waste that is transformed into a useful input for an economic process.* The indicator may be defined on the level of the total waste flow or may be broken down into specific waste flows, e.g. from the various applications of the metals.

Significance of the indicator: The recycling rate of waste materials indicates the potential for further 'closing of cycles' policy.

Method of calculation: The indicator can be calculated in two ways: (1) by determining the fraction of secondary materials in the total input to the economic subsystem, and (2) by determining the fraction of waste materials being reconverted into resources.

Interpretation for environmental policy: The recycling rate may serve as a measure of the efficiency of the economic system as a whole. Recycling is not a goal in itself and sometimes, in the case of an inelastic supply of substances, as is applicable to cadmium (Van der Voet et al., 1994; Guinée et al., 1997) and mercury (Instituut voor Europees Milieubeleid and Environmental Resources Management, 1996), recycling may even be counterproductive. It is therefore difficult to establish a reference value, but in general for substances with an elastic supply one may assume that the higher the recycling rate the better for the environment, assuming leakage from recycling processes is minimised; in that case 100% might serve as a reference value. The indicator may gain more depth when information concerning the nature of the recycling is included. In the case of metals, waste flows may again be transformed into secondary metals, which may then serve as raw materials for all sorts of applications, thereby reducing the need for virgin materials. Waste flows may also be re-used in the shape of waste management residues, such as fly-ash, slag or compost in building, road construction and agriculture. The metals contained in these materials do not serve a purpose but are regarded as contaminants. Nevertheless this may be useful since it prevents the metals from being emitted to the environment, at least for a while, but it does not reduce virgin requirements. In the case of metals, we make this distinction.

11. Use level

The indicator is *the amount of the substance being used in applications within the region.*

Significance of the indicator: The total amount of use within a region determines, directly or indirectly, all other economic flows. It is therefore a measure of the total economic throughput.

Method of calculation: For the use level, a choice must be made for either the total use *flow*, which is the flow of newly produced goods into the use/consumption stage, or the

total *stock* in use. Both can be directly derived from the overview of flows and stocks. It depends on the substance, or rather on the application, which approach makes the most sense.
Interpretation for environmental policy: No reference value can be defined for the level of use. For this indicator, again the 'less is better' adage can be adopted, since in accordance with notions of dematerialisation less use generally means less input, and therefore less emissions, which is beneficial for both resource depletion and pollution. For certain substances, a volume policy can be imagined, which goes hand in hand with the definition of a volume reduction target. Such a target could be applied to the use level, either stock or flow. This indicator has not been applied in the present case study.

12. Economic dissipation
The indicator is *the fraction of a substance's economic flows that is connected with 'trace applications'*.
Significance of the indicator: Dissipative applications are becoming increasingly important as pollution sources. Perhaps the most dissipative application of metals is as a trace element in products. A major proportion of these applications are in fact non-functional: for example, occurrence as contaminants in phosphate fertiliser or fossil fuels. Product policies, such as the Cadmium Directive (Council of the European Communities, 1991), are directed towards such trace applications, functional as well as non-functional. As a result of waste management policies, however, trace applications are on the increase: waste materials such as manure, fly-ash, compost and sewage sludge, in which metals tend to accumulate, are being increasingly re-used. Thus, discarded metals start a second economic life cycle in dissipative applications, as a contaminant of such secondary materials.
Method of calculation: This indicator requires information that does not emerge from the overview of flows and stocks, but is used in the process of quantifying the overview: the breaking down of either the use flow or the use stock into the separate products, and the composition of those products. Based on that information, a classification can be made into trace and bulk applications.
Interpretation for environmental policy: Since trace applications are much more difficult to recycle, and emissions from trace applications are much more difficult to prevent compared with bulk applications, it could be concluded that the lower the share of trace applications, the better it is for the prevention of chain leakages. This indicator has not been applied to the case of heavy metals in the Netherlands, since we did not succeed in operationalising it sufficiently.

13. Disturbance rate

The indicator is the *amount of emissions entering the environment compared to the natural generation of the substance within the region.*
The significance of the indicator: The indicator is designed to provide information on the risk of the natural substance cycle being disturbed by human activity, in the shape of the economic substance cycle. Two ways of approaching this risk are presented below. [** Enkele regels verder zijn er drie???]
Method of calculation: The extent to which the natural cycle is disturbed by human interference can, as a first possibility, be expressed by relating natural, 'virgin' generation of the substance (by formation or erosion) to the anthropogenic addition (by emission). A

second way to provide an indication of the risk of disturbance of the natural cycle is to compare the respective magnitudes of the economic and the ecological inputs. Relatively large economic inputs then constitute a large risk for disturbance of the natural cycle by unwanted or unavoidable losses. A third possibility is to make the comparison on the basis of anthropogenic versus natural *use*.
Interpretation for environmental policy: In all indicator calculation procedures, a higher figure indicates a greater risk of disruption. A standard based on an 'allowable' emission is lacking. A zero disruption risk may serve once more as an ideal, unachievable reference value. The indicator is of interest mainly for the principal biogeochemical cycles (C, N, P, S). For metals, its relevancy may be limited since the anthropogenic contribution will be dominant in any case. It is not applied to the case study in Section III.1 and IV.3.

Of the thirteen indicators presented above, nine have been applied to the case of heavy metals in the Netherlands: all of the indicators for the fate of the mined metals, the human and ecosystem health indicators, and only two of the indicators for the design of a sustainable management regime: the efficiency and the recycling rate indicators. Sections III.1 and IV.3 contain the results.

References

- Adriaanse, A.; S. Bringezu; A. Hammond; Y. Moriguchi; E. Rodenburg; D. Rogich; H. Schütz (1997): Resource Flows: the Material Basis of Industrial Economics. World Resources Institute, Washington DC, 31 pp. + appendices
- Annema, J.A.; E.M. Paardekooper; H. Booij; L.F.C.M. van Oers; E. van der Voet; P.P.A. Mulder (1995): Stofstroomanalyse van zes zware metalen - Gevolgen van autonome ontwikkelingen en maatregelen. RIVM report no. 601014010, Bilthoven, vi + 166 pp. + appendices
- Anonymous (1991): Integrated Substance Chain Management. The Netherlands' Ministry of Housing, Spatial Planning and Environment, Publikatiereeks Produktenbeleid 1991/4, 31 pp. + appendices
- Ayres, R.U. (1989): Industrial Metabolism. In: Ausubel and Sladovich (eds.): Technology and Environment, pp. 23-49. Nat. Academy Press, Washington DC
- Ayres, R.U.; S.R. Rod (1986): Patterns of Pollution in the Hudson-Raritan Basin. Environment 28 pp 14-20 and 39-43
- Azar, C., J. Holmberg and K. Lindgren (1996). Socio-ecological indicators for sustainability. Ecol. Econ., 18: 89-112.
- Brunner, P.H. et al. (1997): Materials Accounting as a Tool for Decision Making in Environmental Policy (MAcTEmPo). 1st Progress Report, April 1997. TU Vienna, Linköping University, ETH Zürich and Leiden University
- Council of the European Communities (1991): 10th Amendment of Directive 76/769/EEC on the approximation of the laws, regulations and administrative provisions of the Member States relating to restrictions on the marketing and use of certain dangerous substances and preparations, informally known as the Cadmium Directive. Brussels, Directive 91/338/EEC, enacted 18-6-1991
- Guinée, J.B.; J.C.J.M. van den Bergh; J. Boelens; P.J.Fraanje; G. Huppes; P.P.A.A.H. Kandelaars; Th.M. Lexmond; S.W. Moolenaar; A.A. Olsthoorn, H.A. Udo de Haes, E. Verkuijlen; E. van der Voet (1998): Evaluation of risks of metal

flows and accumulation in economy and environment. CML working paper, Submitted to Ecological Economics
- Jager, D.T.; C.J.M. Visser (eds.) (1994): Uniform System for the Evaluation of Substances (USES), version 1.0. National Institute of Public Health and Environmental Protection (RIVM), Ministry of Welfare, Health and Cultural Affairs (WVC), Ministry of Housing, Spatial Planning and the Environment (VROM). RIVM Distribution No. 11144/150
- Kleijn, R.; A. Tukker; E. van der Voet (1997): Chlorine in the Netherlands, Part I, An Overview. Journal of Industrial Ecology 1 (1) 95-116
- Kuik, O; H. Verbruggen (1991): In search of indicators for sustainable development. Series Environment & Management Vol. 1, Kluwer Academic Publishers, Dordrecht/Boston/London, vii + 126 pp
- Moolenaar, S.W., S.E.A.T.M. van der Zee; Th.M. Lexmond (1997): Indicators of the sustainability of heavy metal management in agro-ecosystems. The Science of the Total Environment 201 155-169
- Opschoor, J.B.; L. Reijnders (1991): Towards sustainable development indicators. In: Kuik & Verbruggen (1991), pp 7-28
- Organisation for Economic Cupertino and Development (OECD) (1993a): Group on the State of the Environment Workshop on Indicators for Use in Environmental Performance Reviews. Draft Synthesis report, August 1993
- Organisation for Economic Co-operation and Development (OECD) (1993b): Group on the State of the Environment: Environmental Indicators: Basic concepts and terminology, background paper no. 1. In: Proceedings of the Workshop on Indicators for use in Environmental Performance Reviews, 1/2 February 1993, Paris
- Paustenbach, D.J. (ed.) (1989): The Risk Assessment of Environmental and Human Health Hazards: A Textbook of Case Studies. John Wiley & Sons, New York / Chichester / Brisbane / Toronto / Singapore
- Stigliani, W.M.; S. Anderberg (1992): Industrial Metabolism at the Regional Level: the Rhine Basin. IIASA working paper WP-92-10, Laxenburg Austria, 40 pp
- Stigliani, W.; W. Salomons (1993): Our fathers' toxic sins. New Scientist 11 December 1993, pp. 38-42
- Udo de Haes, H.A.; L. Reijnders; H. Verbruggen; L. Hordijk, J.B. Opschoor; F.A.M. de Haan; Th.G. Drupsteen (1992): Accumulation of metals in economic/environmental cycles: mechanisms, risks and possible management strategies. Research program for Sustainability and Environmental Quality, funded by the Dutch Organisation for Scientific Research
- Udo de Haes, H.A.; E. van der Voet; R. Kleijn (1997): Substance Flow Analysis (SFA), an analytical tool for integrated chain management. In: Bringezu, S. et al (eds.): Regional and National Material Flow Accounting: from Paradigm to Practice of Sustainability. Proceedings of the ConAccount workshop 21-23 January 1997, Leiden
- Voet, E. van der, R. Kleijn, L. van Oers; R. Heijungs; R. Huele; P. Mulder (1995): Substance Flows Through the Economy and Environment of a Region. Part I: Systems Definition. Environ. Sci. & Pollut. Res. 2 (2) 90-96
- Voet, E. van der; R. Heijungs; P. Mulder; R. Huele; R. Kleijn; L. van Oers (1995): Substance Flows Through the Economy and Environment of a Region. Part II: Modelling. Environ. Sci. & Pollut. Res. 2 (3) 137-144

- Voet, E. van der; J.B. Guinée, J.B.; H.A. Udo de Haes (1998): Indicators for integrated substance chain management as a measure for environmental quality and sustainable development. CML report no. 139, Leiden, 45 pp
- Wackernagel, M.; W. Rees (1996): Our ecological footprint - reducing human impact on the Earth. The New Catalyst Bioregional Series, New Society Publishers, Gabriola Island, Canada, xi + 108 pp.
- Weizsäcker, E. von; A.B. Lovins; L.H. Lovins (1997): Factor Four - doubling wealth, halving resource use, the new report to the Club of Rome. Earthscan Publications Ltd, London, xxix + 322 pp.

II.7 Combining SFA and economic models

Mathijs Bouman, Reinout Heijungs, Ester van der Voet, Jeroen van den Bergh & Gjalt Huppes

II.7.1 Introduction

Many environmental problems can be directly related to flows of substances, materials and products through the economy. Several methods have been developed to study such flows, but these include no description of economic mechanisms (allocation, optimisation, substitution) or costs and benefits. Economic models, on the other hand, have focused mainly on abstract externalities and do not explicitly describe the flows and transformation of materials. Some form of integration of these two classes of model thus seems desirable.

Such integration has been attempted a number of times. Evidence is provided by references such as Ayres & Kneese (1969), Leontief (1970), Victor (1972) and Perrings (1987). However, none of these attempts has been completely satisfying. The issue at stake is one of conflicting requirements. On the one hand, the models should be complete, in the sense of covering extraction and pollution, production, consumption and waste treatment, bulk materials and micro-pollutants, and so forth. On the other hand, the models should be operational, in the sense of having a low data demand and being easy to construct and run in practice. This second requirement has stimulated the development of a class of rather restricted models, such as substance flow analysis and material flow analysis, life cycle assessment and risk analysis on the physical side, and equilibrium models and macro models on the economic side. These models have modest pretensions in the sense of not aiming to provide an ultimate answer to policy questions. A natural question is then to what extent the results thereby obtained are valid, to what extent expansion of one restricted model with another is feasible and useful, and where the practical boundaries of application and domain extension lie.

There are theoretical surveys of this class of partial models (see, for instance, Kandelaars, 1998). Such overviews usually contain a catalogue of abstract properties, such as primary object and main assumptions. In this section a different approach is taken: that of showing the consequences of the differences between these models in a hypothetical case study. This paper therefore addresses the issue of integration by comparing the approaches and the results of a single case study as can be obtained by means of three partial models: substance flow analysis (SFA), life cycle assessment (LCA) and partial equilibrium analysis (PEA). Clearly, these three are not the only models that are used to study economy-material interactions. There is a wide range of other models, but we feel that the three models discussed here are representative of the typical differences existing between the various model types.

II.7.2 The example

The example system is depicted in Figure II.7.1. This figure describes the relations between 10 different 'nodes' involved in production, use and post-use processing of automobile batteries. As can be seen from Figure 1, stocks of materials or products do not exist in the example. The reason for this assumption is that including stocks would necessitate description of dynamic relations within and between nodes. Including dynamics would reveal differences in the way the models handle time-related issues, but would do so at the expense of a major increase in the complexity of the example. Another important simplification is that in the example materials are the single factor of production. Other factors, such as labour and capital, are excluded from the analysis. It should be noted that the example is not realistic, i.e. the chosen values are not based on data. It is meant as an illustration of the way the three models handle such a system.

Figure II.7.1 Flow diagram of the battery example system.

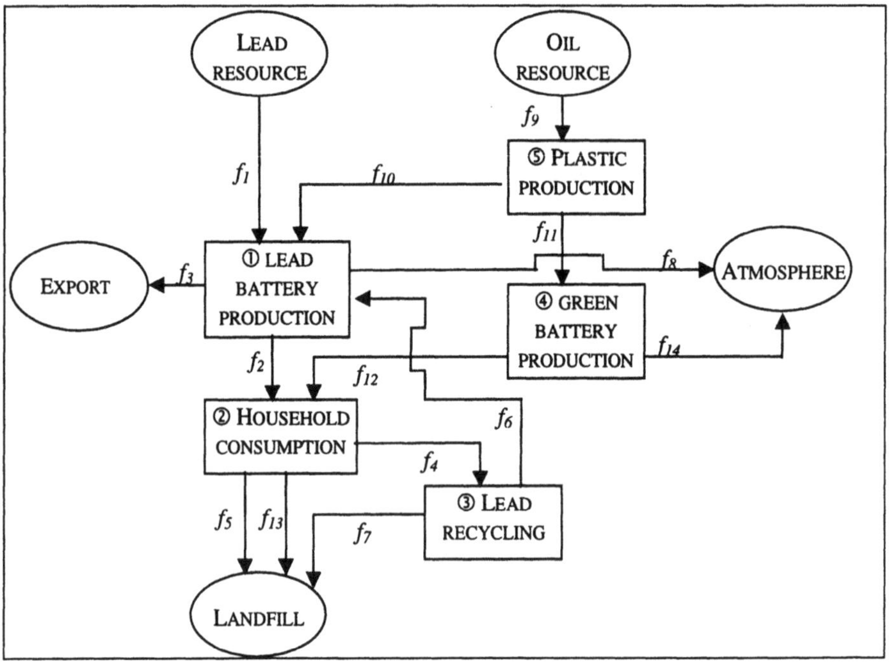

Legend: arrows represent flows, boxes processes and ellipses mines and sinks. The processes are identified by circled numbers.

We assume that batteries come in two types: lead batteries and 'green' batteries. The former is the 'traditional' battery that consists of a lead core and a plastic casing. The latter is its supposedly 'environmentally friendly' substitute that, for sake of simplicity, consists merely of plastics. The initial values of the flows between nodes are shown in Table II.7.1. These values are the starting point of the application of each of the models in the next section. It can be seen that in the initial situation 75% of the batteries are lead batteries. Only one out of every three lead batteries is collected, and from each collected lead battery 80% of the lead is recovered. The bulk of the lead in the economy is dumped. Compared to lead dumping, emissions of lead to the atmosphere are small.

Table II.7.1 Initial values of material flows.

#	Name	Quantity	Unit
f_1	mined lead ore	800	kg/yr
f_2	domestically sold lead battery	150	units/yr
f_3	exported lead battery	45	units/yr
f_4	collected used lead battery	50	units/yr
f_5	dumped used lead battery	100	units/yr
f_6	recycled lead	200	kg/yr
f_7	dumped recycling residuals	55	kg/yr
f_8	air emission, lead battery production	25	kg/yr
f_9	crude oil	75	kg/yr
f_{10}	plastic battery casing	195	units/yr
f_{11}	plastic for green battery	55.5	kg/yr
f_{12}	domestically sold green battery	50	units/yr
f_{13}	dumped used green battery	50	units/yr
f_{14}	air emission, green battery production	5.5	kg/yr

These initial values imply that each lead battery consists of 5 kg lead and 0.1 kg plastic (the battery casing), while a green battery consists of 1 kg plastic. We assume that crude oil, plastic and flue gases contain 1% lead. The initial values have been chosen such that mass balance holds throughout the example.

In our example, production and consumption of batteries are assumed to generate three types of environmental damage: depletion of resources (lead ore and crude oil), air pollution (from lead battery production and from green battery production) and waste dumping (by households and by the recycling sector). Congruent to these three problems, we distinguish three policy objectives for the environmental policy-maker: (1) reduction of the use of virgin materials, (2) abatement of emissions to the atmosphere and (3) reduction of waste dumping.

II.7.3 The models: SFA, LCA and PEA

In this section the three selected concrete models for analysing the relationship between economy and environment — MFA/SFA, LCA and PEA — are discussed separately. In Section II.7.4 they will be applied to the example described above.

Material flow analysis and substance flow analysis (MFA/SFA)
MFA/SFA modelling is based on input-output analysis (IOA), as originally developed by Leontief (1966) and extended in various directions (see Miller & Blair (1985) for a standard reference, and Duchin & Steenge (1998) for a survey of environmental extensions). Input-output analysis is a standard economic tool describing mutual deliveries between sectors, in terms of money or in terms of volumes of goods. It is used mostly on the national level to obtain a picture of the structure of the economy and the mutual relations between economic sectors, and to identify the major flows of money and/or goods within the economic system. It is used as an accounting tool: the

mutual deliveries are 'measured' and summarised in a table, the input-output table. It is also used as a model, i.e. input-output analysis, mainly to predict the changes in sectoral activity as a result of an increase in the final demand for one specific good. This is the so-called impact analysis by means of Leontief multipliers.
An input-output table contains data obtained by observation. Although the data are obviously the result of a complicated mix of behavioural and technical considerations, no attempt is made to explain the data or to separate behaviour from technology. In performing input-output analysis, moreover, the data are treated quite mechanically as technical coefficients. Non-linearities, due for instance to decreasing marginal utility or production, are not considered. Input-output analysis is therefore a rather restricted type of model. In principle it excludes environmental concerns. However, it should be noted that the concept of input-output analysis has been extended by many authors to include environmental aspects; see, for example, Ayres & Kneese (1969), Leontief (1970), Victor (1972), Perrings (1987), Idenburg (1993), Van der Voet (1996), Heijungs (1997).

MFA/SFA modelling, which originates from considering the economy in the physical dimension as described by Ayres (1989) in the concept of industrial metabolism, is rather similar to IOA and is therefore sometimes referred to as 'environmental input-output analysis' (Schrøder, 1996). The mass balance principle is the core rule in MFA/SFA. Applying it rigorously enables one to spot hidden or unexpected flows and emissions, and to detect accumulation of stocks in the economy or the environment, which may cause problems at some future time. Static and steady-state models are used to assess the origins of pollution problems and, in a manner very similar to IOA, to estimate the impacts of certain changes in the economic management of materials (e.g. Baccini and Bader, 1996). Dynamic models are used to estimate the development of emissions and waste generation in future (e.g. Bergbäck and Lohm, 1997). The SFA matrix of coefficients is not drawn up on a sector-by-sector basis, but on a commodity-by-commodity basis. The SFA matrix of coefficients is therefore square, but exponentially larger than the IOA matrix.

Life Cycle Assessment (LCA)
LCA is a tool to assess the environmental consequences of a product from the cradle to the grave. It is intended to support decisions with respect to purchase, improvement, design and so on. LCAs can produce results at the level of the interventions (emissions, extraction of natural resources), at the level of impact categories (global warming, toxicity), at the level of damage to endpoints (human health, material welfare) or at the level of one single indicator. The life cycle of the product generally comprises such diverse aspects as resource extraction, manufacturing of materials and energy, manufacturing of the product, use, maintenance and waste treatment. Capital goods are often only incorporated in as far as their direct functioning is involved (for example, not the depreciation of the truck needed to transport aluminium but only fuel needs and exhaustion gases are included). The procedure for LCA is standardised to some extent: an ISO-standard is under construction, but it will concentrate on procedural matters and main lines of approach, neglecting technical details like formulae.

Main phases of the LCA procedure are:

- goal and scope definition, mainly containing a description of the exact topic, question and approach;

- inventory analysis, concentrating on the physical exchange between product life cycle and the environment in terms of emissions and extractions;
- impact assessment, concentrating on the impacts that can be associated with the aforementioned emissions and extractions;
- interpretation, dealing with uncertainty analyses, preferences, aspects of feasibility and so on.

LCA focuses on the function of a product, not on the product itself. An example of such a function is 'lighting a room with a certain amount of light for 3 hours'. Usage of this so-called functional unit enables a comparison of product alternatives and (re)design of products and/or processes on the basis of the function that is to be delivered by the alternatives. It also implies the study of the so-called product system, from the cradle to the grave. LCA assigns a set of numbers (or one single index) to each alternative fulfilling the specified product function. The numbers only have meaning in a comparative sense. The comparison may be across a range of products fulfilling comparable functions/services (e.g. light bulbs of different types), of a product function within an entire set of product functions (e.g. laces as a part of shoes) or within a product life cycle (e.g. the production stage of the paint within the whole life cycle of cars). Due to uncertainties and assumptions throughout the entire procedure, the outcomes of an LCA should be interpreted with great care, and preferably include extensive sensitivity analyses.

LCA encompasses various types of substances and environmental impacts. In the case of metals, the inventory analysis of all LCAs will include extractions of metal ores, refinement and production of metals, intended and non-intended application in products and intermediates, processing of metal-containing waste, and releases of metals to the atmosphere, to watercourses or to soil. Furthermore, LCAs may be performed for products made of metal or containing it. Aluminium cans and batteries are well-known examples. There are no special requirements for including metals in an LCA. Particular problems that may be encountered are the fact that emissions are often specified in an aggregated form (e.g. 'heavy metals' instead of 'Cd', 'Cr', etc.), and that the chemical specifies of these releases is often not given (e.g. 'Cu' instead of '$CuSO_4$', '$CuCl_2$', etc.). A final remark is that one cannot make an LCA for a metal. Since an LCA is coupled to an application of that metal, many LCAs of metal-containing products may be conducted. On the other hand, all these products are associated with non-metallic substances, so that the LCA of a metal-containing product contains information on flows of sulphur, carbon and may other substances.

LCA modelling, comparable to SFA modelling, is based on IOA: it describes deliveries from one process to another using only linear relationships. Contrary to SFA, it takes only a limited part of the economy into account, and in a much more detailed manner. Its matrix of coefficients is based on the single output: processes are taken into account only if and to the extent that they contribute to the functional unit.

Partial Equilibrium Analysis (PEA)

Partial equilibrium models describe the outcome of a market or a set of markets by depicting the behavioural relations that underlie the outcome. This means that the impact of a change in, for instance, environmental policy can be tracked down to its effects on consumption and production decisions. Since the decision rules are explicitly modelled, the price effects and substitution effects of a given policy can be analysed. The results of PEA depend heavily on the assumption that all market actors maximise their pay-off by equating marginal benefits and marginal costs, and the assumption that all markets are in equilibrium (see Cropper and Oates (1992) for a survey of economic equilibrium models of environmental problems, and Baumol and Oates (1988) for *the* classic introduction in this field).

Partial equilibrium modelling may contain various types of equations, as well as inequalities. Linear equations may be used to enforce mass balance, for example, as described in Section II.2. Non-linear relations are used to describe production or demand functions. It is an extensive task to determine the variables to fit the situation as depicted in Figure II.7.1. This can best be shown by application: see Section II.7.4 below.

II.7.4 Application of the models to the example

The SFA model

For the SFA model, the data of Figure II.7.1 are translated into IO-like equations. Letting y-variables represent the amount of lead contained in the f-flows, the set of equations contains exogenously fixed variables of the type $y_1 = a$, dependency equations of the type $y_2 = b * y_1$ and balancing equations like $y_3 = y_2 - y_1$.

Exogenously determined variables are domestic demand for lead batteries (y_2), domestic demand for green batteries (y_{12}), total production of lead batteries ($y_2 + y_3$) and the matching total production of plastic casings for the lead batteries (y_{10}). See Table II.7.2 for an explanation of the variables and coefficients.

$$y_2 = f \times (a + c \times b)$$
$$y_{12} = g \times (a + c \times b)$$
$$y_2 + y_3 = h \times (a + c \times b)$$
$$y_{10} = h \times c \times b$$

Dependency equations are formulated for the emissions to the atmosphere from both lead battery production (y_8) and green battery production (y_{14}), for the collection of discarded lead batteries (y_4), for the recovery of secondary lead from the collected lead batteries (y_6) and finally for the dumping of discarded green batteries (y_{13}):

$$y_8 = i \times (y_1 + y_6)$$
$$y_{14} = j \times y_{11}$$
$$y_4 = k \times y_2$$
$$y_6 = l \times y_4$$
$$y_{13} = m \times y_{12}$$

This set is completed by so-called balancing equations to calculate the remaining lead flows, at the same time enforcing mass balance. In this way y_1 is calculated, the required amount of freshly mined lead, as well as y_5, the amount of lead batteries being discarded by consumers. The assumption here is that battery consumption is in a steady state and consequently there is no stock change. In this respect the example has its shortcomings: signalling and modelling stock changes is an important part of SFA. Also y_7 (i.e. the amount of unrecovered lead ending up at the landfill site after all), y_9 (i.e. the demand for crude oil in terms of its lead contamination) and finally y_{11} (i.e. the required amount of plastic for the production of green batteries) are calculated by balancing equations.

$$y_1 = y_2 + y_3 + y_8 - y_6 - y_{10}$$

$$y_5 = y_2 + y_{12} - y_4 - y_{13}$$

$$y_7 = y_4 - y_6$$

$$y_9 = y_{10} + y_{11}$$

$$y_{11} = y_{12} + y_{14}$$

Table II.7.2 — Variables and Coefficients

variable/ coefficient	represents	unit	initial value
a	amount of lead in 1 lead battery	kg	5
b	weight of 1 plastic battery case	kg	0.1
c	lead content of plastic	kg/kg	0.01
d	weight of 1 plastic green battery	kg	1
e	total internal demand for batteries	units	200
f	internal demand for lead batteries	units	150
g	total demand for and production of green batteries	units	50
h	total production of lead batteries; total number of battery cases produced	units	195
i	emission coefficient, lead battery industry	kg/kg	0.025
j	emission coefficient, green battery industry	kg/kg	0.0991
k	fraction of discarded lead batteries collected for recycling		1/3
l	fraction of lead recovered from collected batteries		0.7998*
m	fraction of discarded green batteries landfilled		1

*0.8 of lead in batteries; lead from plastic casing is not recovered.

The three policy objectives described in Section II.7.2 are translated into fairly extreme 'policy packages' in order to explore the potential usefulness of such directions:

(i) As a possibility to reduce virgin lead extraction, a complete substitution of lead batteries by green batteries. The absence of economic mechanisms in the SFA model compels us to specify two extremes for the development of lead battery production: (ia), production of lead batteries remains at the same level, batteries are exported, and (ib), production of lead batteries is discontinued altogether.

(ii) In order to reduce lead emissions to air, end-of-the-pipe emission reduction by technical means to 1% of the present level is assumed, with no influence on supply and demand of lead batteries or green batteries.

(iii) In order to prevent landfill, collection of discarded batteries is boosted to 100% and transformation of old batteries into secondary lead to 90%.

This leads to the following changes in coefficients:

Table II.7.3 Changes in variables to simulate the policy packages.

affected variable	initial value	value, package (ia)	value, package (ib)	value, package (ii)	value, package (iii)
f	150	0	0	150	150
h	195	195	0	195	195
g	50	200	200	50	50
i	0.025	0.025	0.025	0.00025	0.025
j	0.0991	0.0991	0.0991	0.000991	0.0991
k	0.333	0.333	0.333	0.333	1
l	0.7998	0.7998	0.7998	0.7998	0.8998

The results of these changes in terms of the flows of lead are presented in Table II.7.9 below.

The LCA model

The data of the flows of products and materials have been manipulated into standard LCA process data according to the normal procedures. These are:

- data are usually normalised to an arbitrary but round output quantity (like 1000 batteries instead of 150 batteries per year); to safeguard transparent comparison with SFA and PEA this optional step has not been carried out;
- inputs of flows are indicated by negative numbers, outputs by positive numbers;
- the order of the flows must be changed into one set of economic flows (which flow from or to other processes) and one set of environmental flows (which flow from or to the environment).
- multiple processes (e.g. joint production, waste treatment including recycling), must be split4 [** bij mij ziet dit voetnootnr uit als een gewone '1'] into independent single processes; this applies to process ③ where a so-called allocation factor (λ) is used to allocate the recycling residuals to treatment of collected used lead batteries (process ③a) and production of recycled lead (process ③b) and to process ⑤, where a possibly different allocation factor (μ) is used to allocate the crude oil over production of plastic casings for lead batteries (process ⑤a) and production of plastic for green batteries (process ⑤b);

4 This procedure of splitting a multiple process into two (or more) single processes is referred to in LCA circles as the allocation step. This term may be somewhat confusing for economists, as it may wrongly suggest the incorporation of market allocation mechanisms into LCA.

- the consumption process is separated into consumption of lead batteries and consumption of green batteries;
- the function of the consumption processes needs to be specified; this enters the table as flows a_7 and a_8.

This leads to Table II.7.4. One point needs clarification. LCA studies the material flows associated with a functional unit of product. The calculated flows do not therefore represent the total flows in the economy-environment system. For this reason, the calculations are all done in terms of different symbols (*a* and *b* instead of *f*).

Table II.7.4. Table of process data for LCA.

Flow	Meaning	①	②a	②b	③a	③b	④	⑤a	⑤b	Unit
a_1	total sold lead batteries	$-f_2+f_3-f_2$		0	0	0	0	0	0	units
a_2	collected used lead batteries	0	f_4	0	$-f_4$	0	0	0	0	units
a_3	recycled lead	$-f_6$	0	0	0	f_6	0	0	0	kg
a_4	plastic lead battery casings	$-f_{10}$	0	0	0	0	0	f_{10}	0	units
a_5	plastic	0	0	0	0	0	$-f_{11}$	0	f_{11}	kg
a_6	total sold green batteries	0	0	$-f_{12}$	0	0	f_{12}	0	0	units
a_7	lead battery use	0	g_1	0	0	0	0	0	0	yr
a_8	green battery use	0	0	g_2	0	0	0	0	0	yr
b_1	lead ore	$-f_1$	0	0	0	0	0	0	0	kg
b_2	dumped used lead batteries	0	f_5	0	0	0	0	0	0	units
b_3	recycling residuals	0	0	0	λf_7	$(1-\lambda)f_7$	0	0	0	kg
b_4	air emission, lead battery production	f_8	0	0	0	0	0	0	0	kg
b_5	crude oil	0	0	0	0	0	0	$-\mu f_9$	$-(1-\mu)f_9$	kg
b_6	dumped used green batteries	0	0	f_{13}	0	0	0	0	0	units
b_7	air emission, green battery production	0	0	0	0	0	f_{14}	0	0	kg

We need to choose the allocation factors λ and μ; this is done according to Table II.7.5.

Table II.7.5. Choice of coefficients in LCA

Coeff.	Meaning	Value	Unit
λ	for allocation of process ③ to independent processes ③a and ③b	0.5	-
μ	for allocation of process ⑤ to independent processes ⑤a and ⑤b	0.5	-

The standard theory of LCA now provides a procedure to partition the table of process data into two matrices and to calculate a list of environmental flows associated with a certain unit of function. Here we choose to do calculations for 100 years of lead battery use and 100 years of green battery use. However, to enable a comparison with the other two models, we may translate the matrix equation into a set of simultaneous equations. This requires the explicit introduction of 8 scaling parameters *s*, one for each (single) process. For the case of lead batteries, the set of equations is written below.

For green batteries, we only need to exchange the right-hand side parameters 100 and 0 in the 7th and 8th equation. Solving the equations yields the tabulated results for lead and green batteries respectively. Furthermore, we will be assuming that a weighting between different pollutants and resources has been set, involving weighting factors. Multiplying the environmental results with the weighting factors delivers the weighted impacts for lead and green batteries, which can be found in Table II.7.6 below. We thus see that for the fulfilment of an identical function (100 years of battery use), the two alternatives products have quite different environmental flows and impacts. The lead battery system of course has many lead-related flows and impacts, but oil depletion, in particular, means that the green battery alternative has serious disadvantages (due to our fictitious set of weighting factors).

$$s_1(f_2+f_3)+s_{2a}(-f_2)=0$$
$$s_{2a}f_4+s_{3a}(f_4)=0$$
$$s_1(-f_6)+s_{3b}f_6=0$$
$$s_1(-f_{10})+s_{5a}f_{10}=0$$
$$s_4(-f_{11})+s_{5b}f_{11}=0$$
$$s_{2b}(-f_{12})+s_4f_{12}=0$$
$$s_{2a}a_7=100$$
$$s_{2b}a_8=0$$
$$b_1=s_1(-f_1)$$
$$b_2=s_{2a}f_5$$
$$b_3=s_{3a}\lambda f_7+s_{3b}(1-\lambda)f_7$$
$$b_4=s_1f_8$$
$$b_5=s_{5a}(-\mu f_9)+s_{5b}(-(1-\mu)f_9)$$
$$b_6=s_{2b}f_{13}$$
$$b_7=s_4f_{14}$$

The hypothetical measures that were formulated in the previous subsection have been analysed with LCA once more. Package (i), the take-over of green batteries, is not interesting with LCA, as LCA does not deal with actual market volumes, but just compares lead and green batteries on the functional level, i.e. per year of use.[5] Package (ii), the end-of-pipe reduction of all air emissions by 99% results in a simple calculation: the life-cycle air emissions due to production of lead batteries (f_8) and production of green batteries (f_{14}) have indeed been reduced by a factor of 0.99. Package (iii), increased collection of lead batteries to 100% and their recycling to 90% produces less trivial results. First we must change certain coefficients of the equations, to account for the changes in technology structure. We change the coefficient for output

5 Recall that advantages or disadvantages that are related to scale are outside the linear homogeneous formalism, and hence outside the scope of LCA.

Table II.7.6. Environmental flows according to LCA of 100 years battery use, attached weights and weighted scores for lead batteries and green batteries.

Flow	Meaning	Environmental flows		Weighting factor	Weighted impacts	
		Lead batteries	Green batteries		Lead batteries	Green batteries
b1	lead ore	-82 (kg)	0 (kg)	–5 (kg^{-1})	410	0
b2	dumped used lead batteries	13 (units)	0 (units)	20 (per unit)	267	0
b3	recycling residuals	6.5 (kg)	0 (kg)	3 (kg^{-1})	19	0
b4	air emission, lead battery production	2.6 (kg)	0 (kg)	25 (kg^{-1})	64	0
b5	crude oil	-3.9 (kg)	-15 (kg)	–40 (kg^{-1})	154	600
b6	dumped used green batteries	0 (units)	20 (units)	5 (per unit)	0	100
b7	air emission, green battery production	0 (kg)	2.2 (kg)	2 (kg^{-1})	0	4
	weighted TOTAL				914	704

of collected used batteries by process ②a from 50 to 150, for output of dumped used lead batteries by that process from 100 to 0, for output of recycled lead by process ③b from 200 to 229.5 and for output of recycling residuals by processes ③a and ③b from 27.5 to 12.25. The amount of dumped used lead batteries (f_5) then drops from 20 to 0 and the amount of recycling residuals (f_7) drops from 6.5 kg to 6 kg. For the green batteries there is of course no difference. The results are shown, along with that of the other models, in Table II.7.9.

The PEA model

In the example transfers take place on five different markets: a market for lead (both new and recycled), for oil, for plastics, and a domestic and a foreign battery market. The model presented below accounts for four markets, because we exclude the oil market. We assume that the raw and intermediate material markets (for lead and plastics) are international markets characterised by perfect competition. This implies that the prices of lead and plastics are determined on the world market. On the battery market firms do have some monopolistic leverage, so they can to a certain extent determine the prices of their output. The functional form of the model is described below.

<u>Lead battery production and consumption</u>

Ignoring the plastic casings, the inputs for lead battery production are new lead and recycled lead. The production function reads:

$$(x_2 + x_3) = \gamma (x_1 + x_6)^{\alpha}, \quad \gamma > 0, \quad 0 < \alpha < 1, \tag{1}$$

which describes a decreasing returns to scale technology (i.e. the average amount of lead required to produce one lead battery rises with the level of production). Equation (1) implies that new lead and recycled lead are perfect substitutes. Therefore, demand for each input is infinitely elastic, so for non-zero x_1 and x_6 the market price of new lead and recycled lead are identical:

$$p_1 = p_6 \, . \tag{2}$$

The inverse domestic demand function for lead battery is given by

$$p_2 = \beta (x_2)^{\mu} (x_{12})^{\sigma}, \quad \beta > 0, \quad -1 < \mu < 0, \quad -1 < \sigma < 0, \quad \mu < \sigma \tag{3}$$

For simplicity, we assume that export of lead batteries is a fixed fraction π of total lead battery production, or

$$x_3 = \pi (x_2 + x_3), \quad 0 > \pi > 1 \, . \tag{4}$$

Green battery production and consumption

Plastic is the single input in production of green batteries, so

$$x_{12} = \varepsilon (x_{11})^{\rho}, \quad \varepsilon > 0, \quad 0 < \rho < 1, \tag{5}$$

describes the decreasing returns technology of firms in the green battery producing sector. The inverse demand function for green batteries is

$$p_{12} = \omega (x_{12})^{\xi} (x_2)^{\sigma}, \quad \omega > 0, \quad -1 < \xi < 0, \quad \xi < \sigma \, . \tag{6}$$

The restrictions on σ, μ, and ξ in equation (3) and (6) guarantee that lead batteries and green batteries are (imperfect) substitutes, and that the cross-price elasticity is smaller than the own-price elasticities.

Recycling

We assume that purely economic motives play no role in the collection of used lead batteries. The collection rate (λ) is therefore exogenously determined:

$$x_4 = \lambda x_2, \quad 0 \geq \lambda \geq 1 \, . \tag{7}$$

Lead is recovered from the collected lead batteries using a decreasing returns recycling technology that can be described by an exponential function:

$$x_6 = \delta x_4 \left(1 - e^{-S}\right), \quad \delta > 0, \quad S > 0, \tag{8}$$

where S is the level of recycling activity and δ is the *ex post* lead content of a single lead battery. Denoting air emissions of lead per lead battery by ν, the amount of lead per battery is

$$\delta = \frac{(1 - \nu)(f_1 + f_6)}{f_2 + f_3}, \quad 0 < \nu < 1, \tag{9}$$

Owing to the decreasing returns production function (1), δ changes with the level of lead battery production.

Pollution

In this model there are five sources of pollution. Dumping of used lead batteries (f_5) is given by the difference between used and recovered batteries,

$$f_5 = f_2 - f_4. \quad (10)$$

Dumping of lead by the recycling sector (f_7) is the difference between the lead contained in recovered batteries an the amount of recycled lead,

$$f_7 = \delta f_4 - f_6. \quad (11)$$

Assuming that lead emissions from lead battery production can (partly) be avoided by implementation of abatement technology, air pollution generated by production of batteries (f_8) is gross air pollution minus abated pollution (*B*),

$$f_8 = \nu(f_1 + f_6) - B \quad (12)$$

where the abatement technology is such that for all levels of lead-emission abatement activity (*A*):

$$B = \psi A^{\theta} \quad (13)$$

The price of *A* is normalised to unity. Since green batteries are not recovered, dumping of green batteries (f_{13}) is

$$f_{13} = f_{12}, \quad (14)$$

Unintentional lead emissions from green battery production (f_{14}) are given by

$$f_{14} = \phi f_{11}, \quad 0 < \phi < 1, \quad (15)$$

where ϕ is the amount of lead emitted per unit of green battery produced. Note that the description of the pollution flows (equation 10-15) implies that mass balance holds.

Calibration

Assuming that all firms maximise profits and all markets are in equilibrium, the model can be explicitly solved for all endogenous variables. The parameters and exogenous prices are chosen such that the initial numerical solution of the model is in accordance with the values in Table II.7.1. These parameter values and prices are shown in Table II.7.7 below. Note that the values of β and ω, and the values of μ and ξ are the same. This means that for identical prices of lead batteries and green batteries, demand for each type battery is the same.

Table II.7.7 Parameter values and exogenous prices.

parameters				exogenous prices	
α	0.3	ω	1447.5	$p_1 = p_6$	1
β	1447.5	ε	15.5	p_{11}	8.8
γ	18.9	ξ	-0.5	r	49.9
μ	-0.5	ν	0.025		
σ	-0.25	ϕ	0.0011		
λ	0.33	ψ	1		
ρ	0.3	θ	0.5		
π	0.23				

This model is then used to analyse the policy options for attaining the three aforementioned environmental goals (i.e. reduction of the use of new lead, reduction of air pollution and reduction of waste dumping). Since the general conclusion in economic theory is that the most efficient mode of environmental policy is generally one that uses taxes to change the behaviour of economic agents, we will focus mainly on tax instruments. The reduction options mentioned in Section II.7.2 are translated for PEA application as follows:

(i) reduction of resource depletion by taxing virgin lead and by subsidising green batteries
(ii) abatement of air pollution by taxing emissions
(iii) reduction of landfill by promoting lead battery collection and subsidising recycling.

This is simulated in the model by making the changes as described in Table II.7.8 below:

Table II.7.8 Simulation of the three reduction options in the PEA model.

package	(ia)	(ii)	(iiia)	(iiib)	(ib)
modification	100% tax on new lead tax_1=1	tax on air emissions tax_8=9.5	100% collection λ=1	full subsidy on recycling tax_6= -125	subsidy on green battery tax_{12}=-0.95

The results are presented in Table II.7.9, together with the results of the calculations with the SFA and the LCA model.

Results of the calculations of the three models

In Table II.7.9 below, the results of the calculations of the three applied models are presented. As can be seen, they differ in various aspects although the same general measures have been implemented. Due to the inherent characteristics of the models, treatment of the various abatement options has been quite different. This is discussed below.

If we regard the options for **reducing the virgin input of lead**, we see that the SFA and LCA model zoom in on the technical solutions. Both generate the option of substituting lead batteries by green batteries. The PEA model addresses not technical measures but (economic) instruments: a tax on virgin lead, an emission tax and a subsidy on green batteries are introduced. Here we see not so much a contradiction, but a different level of entrance into the realm of problem-solving options. Looking at the results we see that the SFA results can be compared quite well with the PEA results. SFA tells us that substitution, if implemented in its extreme form, is very effective in solving the depletion problem. Apart from that it also solves the waste and the emissions problem. The PEA options can be regarded as instruments to implement such a substitution. The effectiveness of both options is of course less in the PEA model, and is somewhere in between the baseline and the extreme SFA package (ib); we also see that in this case the tax on virgin lead is more effective than the subsidy on green batteries. The question of what happens to the lead battery production sector is addressed differently, exogenously and inadequately in both models.

Table II.7.9 - Effects of the three policy packages as calculated by the three models.

	baseline		(i) substitution		(ii) air emission reduction		(iii) recycling	
SFA model results, total flows of lead, steady state			*export LBs*	*close LB prod.*	*99% emission reduction*		*100% collection, 90% recycling*	
required virgin lead (kg/y)	800		1000	0	775		325	
emissions of lead to atmosphere (kg/y)	25		25	0.22	0.25		25	
landfill lead (kg/y)	551		2	2	551		75.5	
PEA model results, total flows of lead, steady state			*tax lead*	*subs. GB*	*tax air emissions*		*boost collect*	*subsid. recycl.*
required virgin lead (kg/y)	800		359	598	633		300	750
emissions of lead to atmosphere (kg/y)	25.06		11.06	20.22	0.46		25.06	25.06
landfill of lead (kg/y)	550.5		245.0	433.4	439.0		50.6	500.6
price of lead batteries ($unit)	44		50	36	46		44	44
price of green batteries ($/unit)	59		62	703	59		59	59
LCA model results, emissions / extractions per battery	*LB*	*GB*	**n.a**		*LB 99% emiss. red.*	*GB 99% emiss. red.*	*LB 100% collect ., 90% rec.*	*GB*
required virgin lead (kg)	-82	0			-82	0	-82	0
dumped used LB (units)	13	0			13	0	0	0
recycling residuals (kg)	6.5	0			6.5	0	6	0
air emission, LB production (kg)	2.6	0			0.026	0	2.6	0
required crude oil (kg)	-3.9	-15			-3.9	-15	-3.9	-15
dumped used GB (units)	0	20			0	20	0	20
air emission, GB production (kg)	0	2.2			0	0.022	0	2.2
weighted total	*914*	*704*			*851*	*700*	*646*	*704*

The SFA model regards two extremes, which influences the results quite substantially: only in the drastic package (ib) is the problem solved, while in package (ia) it even increases. The PEA model assumes foreign demand to be influenced in the same way as domestic demand. The LCA result is the comparison between green batteries and lead batteries and tells us whether these green batteries are indeed preferable from an environmental point of view. It turns out that a substitution would cause a shift from lead depletion and emissions to oil depletion and hydrocarbon emissions. In all, the results of the three models are not contradictory but complementary.

The second option is the **reduction of air emissions**. SFA assesses this option by assuming 99% effective filters in place, and PEA by introducing an emission tax, which again can be viewed as a policy instrument used to implement the supposed techniques. SFA and PEA results point in the same direction again, but the PEA model shows that there are some economic consequences of this tax which are ignored in the SFA model: the demand for lead batteries drops slightly, causing also the required virgin input and the landfilled waste to decrease. In this case, the PEA model seems to encompass the SFA model and adds to it. The LCA model has no additional value here.

The third option is to **decrease landfill of lead-containing waste**. Both the SFA and the PEA model aim to do this by boosting collection and recycling, SFA by assuming that it happens and PEA by introducing a subsidy on recycling. Here the results differ somewhat. Apart from the fact that this subsidy apparently is not very effective in boosting recycling, we also see that the effectiveness from the point of view of the landfill problem is virtually zero in the PEA model, while it is quite substantial in the SFA model. This is due to the fact that the subsidy is not assumed to influence collection but only recovery in the PEA model, which is raised to the extreme of 100% in the SFA model. Again the two models appear to be complementary rather than contradictory: SFA tells us that in principle recycling may help considerably in solving the problem, while PEA adds that implementation will probably be problematical.

II.7.5 Evaluation of the applied models

What can we learn from the application of the three different models to one and the same example? Conclusions can be drawn at various levels. In the first place, it can be concluded that each of the models serves its own purposes and therefore has its own strong points as well as its own limitations. From the application to the example, it appears that the results of the three models are in most cases complementary rather than contradictory. SFA can be used to assess whether certain options, such as technical measures, might solve the problem in principle. LCA can be used to assess whether certain solutions do not lead to other, also serious environmental problems. PEA can then be used to look for the most efficient mode of implementation, identifying some routes (tax on air emissions) as surprisingly beneficial and others (boosting recycling) as difficult to implement.

In the second place, it appears that this example is indeed a very small and simple one for the two physical models, SFA and LCA, while it is a rather large and complex system for the PEA model. SFA and LCA models usually handle much larger systems, even in theoretical applications. SFA mostly operates at a macro-level, encompassing all economic sectors insofar as they handle the substance involved. LCA is primarily a micro-level tool; the LCA system is large because of the inclusion of processes in a detailed manner and the allocation of tiny parts of macro-level sectors such as energy or transport. Large systems are possible because the modelling equations used for LCA and SFA are all simple linear equations, while the PEA equations are much more complicated. The physical models appear to aim at completeness and obtain their added value from quantity. The PEA model on the other hand, which also operates at the micro-level, aims at much more careful modelling of a few important mechanisms while ignoring the remainder, thus focusing on quality rather than quantity.

A third conclusion following from the above is that both the physical models and the economic model obtain their strength from the observation of mechanisms rather than from describing 'the real world'. The SFA model identifies problem-causing mechanisms based on mass conservation, such as stock-building, creating cycles, poisoning of cycles and connections. The LCA model identifies the main problematical parts of functional chains, options to improve chains and problem-shifting between environmental problems. The PEA model identifies the market mechanisms that can be used most suitably to achieve a certain end, such as welfare optimisation All such mechanisms are relevant and interesting to model, although it is certainly difficult to model them all at the same time.

This leads to some considerations regarding the use of these models. A first and rather straightforward recommendation is not to use the models for purposes they were not designed for. This may seem rather trivial; however, in practice we may observe that this rule is violated in many cases. Other recommendations, such as stated below, refer to the future development and use of economic-environmental models.

II.7.6 Towards Integration

One possible direction for development could be to design a procedure to use such models in tandem, thus using the strong points of each while catching out the respective limitations. If we stick to the example of heavy metals, we could imagine a procedure as follows:

- first use SFA to identify the metal flows, distinguish the problematical flows, select the main flows to regulate and try out the problem-solving potential of some technically defined options
- then use LCA to evaluate the emerging alternatives (either products, materials or production processes) with respect to shifting to other environmental problems
- then use PEA to model the markets connected with the selected flows-to-regulate and evaluate the various possible instruments on their environmental as well as economic consequences
- finally introduce the results for the most promising options out of the PEA model once again into the SFA model to identify unexpected problem-shifting to other parts of the substance chain.

In this way, all models have their proper sequential place without transgressing beyond their natural boundaries, at the same time supporting the evaluation much more strongly together than alone. Theoretically this may be the easiest way to proceed. In practice, this would imply close co-operation between disciplines, which may not be easy but could certainly be worthwhile.

Quite a different direction of thinking is to attempt an integration of the modelling principles of the three models, in order to develop new models that have all the advantages and none of the drawbacks. Models in which economic and physical modelling are integrated already exist: in Sections II.3 and III.3, MPC models are described, adding mass balance equations to a modelling of markets on a micro-level, rather similar to the PEA model in this section. On the macro-level the MARKAL model (Gielen & Kram, 1997) adds one or two markets to a large input-output-like physical structure. Such modelling may also be very valuable and might be extended in other directions to create a new class of integrated economic-environmental models. The main danger of progressing in this direction is falling into the trap of trying to design 'the ultimate model' that can do everything at the same time. In practice it may well be that through integration some of the specific assets of the specialist models are lost. On the other hand, the aforementioned examples of MPC and Markal can be seen as practical compromises in this vein.

Which of these two routes is the most useful, and how to proceed accordingly, cannot be decided on the basis of the present exercise. For the moment it would seem useful to try both. It may depend on the specific question that needs answering. It may even be a matter of taste. In any case this seems to be a field of research that is still wide open for the future.

References

- Ayres, R.U. and A.V. Kneese (1969). 'Production, Consumption and Externalities', American Economic Review, **59**, 282-297.
- Baccini, P. and H.-P. Bader (1996). Regionaler Stoffhaushalt. Erfassung, Bewertung und Steuerung. Spektrum Akademischer Verlag, Heidelberg.
- Bergbäck, B. and U. Lohm (1997). 'Metals in Society'. In: Brune, Chapman, Gwynne & Pacyna (eds), The Global Environment. Scandinavian Science Publishers, Wiley, VCH.
- Bringezu, S., M. Fischer-Kowalski, R. Kleijn & V. Palm (eds) (1997). Regional and National Material Flow Accounting. From Paradigm to Practice of Sustainability. Proceedings of the ConAccount workshop 21-23 January 1997, Leiden, the Netherlands. Wuppertal Institute for Climate, Environment & Energy, Wuppertal Special 4.
- Cropper, M.L. and W.E. Oates (1992). 'Environmental Economics: a Survey' Journal of Economic Literature, **30**, 675-740.
- Duchin, F. and A.E. Steenge, (1998?) 'Input-Output Analysis, Technology and the Environment'. In: J.C.J.M. van den Bergh (ed), Handbook of Environmental and Resource Economics. Edward Elgar, Cheltenham, forthcoming.
- Gielen, D.J. & T. Kram (1997). The MARKAL Model for Environmental Accounting in Energy and Materials Systems. In Bringezu, S. et al. (eds) (1997), see above.
- Guinée, J.B., J.C.J.M. van den Bergh, J. Boelens, P.J. Fraanje, G. Huppes, P.P.A.A.H. Kandelaars, Th.M. Lexmond, S.W. Moolenaar, A.A. Olsthoorn, H.A. Udo de Haes, E. Verkuijlen and E. van der Voet (1998?). 'Evaluation of Risks of Metal Flows and Accumulation in Economy and Environment'. Ecological Economics, forthcoming.

- Heijungs, R. (1997). Economic Drama and the Environmental Stage - Formal Derivation of Algorithmic Tools for Environmental Analysis and Decision-Support from a Unified Epistemological Principle, Centre of Environmental Science, Leiden, the Netherlands.
- Idenburg, A.M. (1993). Gearing Production Models to Ecological Economic Analysis: a Case Study within the Input-Output Framework, for Fuels for Road Transport, University Twente, Enschede, the Netherlands.
- Kandelaars, Patricia P.A.A.H (1998). Material-Product Chains: Economic Models and Applications, Thesis Publishers, Amsterdam.
- Leontief, W. (1966). Input-Output Economics, Oxford University Press, New York.
- Leontief, W. (1970). 'Environmental Repercussions and the Economic Structure: an Input-Output Approach', Review of Economics and Statistics, **52**, 262-271.
- Mackay, D.(1991). Multimedia Environmental Models. The Fugacity Approach. Lewis Publishers, Chelsea.
- Miller, R.E. and P.D. Blair (1985). Input-Output Analysis - Foundations and Extensions, Prentice Hall, Englewood Cliffs NJ.
- Perrings, C. (1987). Economy and Environment - a Theoretical Essay on the interdependence of Economic and Environmental Systems, Cambridge University Press, Cambridge.
- Van der Voet, E. (1996). Substances from Cradle to Grave, Optima Druk, Molenaarsgraaf, the Netherlands.
- Victor, P.A. (1972), Pollution: Economy and Environment, Edgar Elgar, Cheltenham.

Part III Applications of the developed models

Contents:

In Part III, the models described in Part II are applied to the case of heavy metals. In Section III.1 a combination of FLUX, Dynabox and the indicators is applied to the flows and stocks of the four heavy metals in the Netherlands. In Section III.2 a number of applications of MPC modelling are described, each covering a specific materials-product chain from the metals' metabolism. Section III.3, finally, is dedicated to applications of the D(SC)B model to heavy-metal flows in agricultural soils under various circumstances regarding soil composition and agricultural practice.

III.1 Metals in the Netherlands: application of FLUX, Dynabox and the indicators

Ester van der Voet, Jeroen B. Guinée & Helias A. Udo de Haes[6]

III.1.1 Introduction

In this section an application is presented of the combined models FLUX (II.1) and Dynabox (II.5) and the indicators presented in II.5. This modelling-and-evaluation combination is applied to the metabolism of cadmium, copper, lead and zinc for the total Dutch economy (Guinée et al., 1998). To this end an inventory was made of the flows in 1990; FLUX was used to balance the data. Input data for 1990 such as data on flows in the economy, accumulations, emissions and transboundary pollution have been taken mainly from Annema et al. (1995). The 1990 emissions were introduced in Dynabox to calculate the concentrations in the respective environmental compartments and human intake by the various routes.

The 1990 situation is compared with the steady-state situation: the situation that ultimately emerges with indefinite continuation of the present metals management regime. The purpose of this exercise was to gain an insight into the long-term consequences of the current management regime. As argued in Section I.2, it is at present unclear whether the emission reduction realised over the past few decades might not have unforeseen adverse consequences in the future. The procedure for calculating the steady-state situation has been described in Section II.1. In summary, it boils down to establishing the equilibrium associated with the present structure of supply and demand of the metals concerned. Note that the steady state must not be interpreted as a prediction of any future situation: it contains no trends, no socio-economic developments and no policies. It is a tool to evaluate the present management regime with respect to its potential future consequences from a sustainability point of view. The steady-state emissions, the result of FLUX, were introduced in Dynabox once again to obtain the steady-state environmental concentrations.

The results of these calculations with FLUX and Dynabox are presented in Section III.1.2 and evaluated using the indicators for environmental risks and societal metabolism, as described in Section II.6. Most of the indicator values have been calculated by simple spreadsheet manipulations of FLUX results. The indicators for human and ecosystem health risk came from Dynabox. Below, the results of the calculations are discussed. As mentioned in II.6, the indicators are divided into three groups, related to the research questions of the 'Metals' programme:

- indicators for the fate of the mined metals (III.1.3)

[6] This chapter is based on an article published in Ecological Economics: Guinée, J.B., J.C.J.M. van den Bergh, J. Boelens, P.J. Fraanje, G. Huppes, P.P.A.A.H. Kandelaars, Th.M. Lexmond, S.W. Moolenaar, A.A. Olsthoorn, H.A. Udo de Haes, E. Verkuijlen & E. van der Voet (1999). Evaluation of risks of metal flows and accumulation in economy and environment. Ecological Economics 30 (1999) pp 47 - 65.

- indicators for the evaluation of present management in terms of sustainability (III.1.4)
- indicators for the design of a sustainable management regime (III.1.5).

III.1.2 The 1990 and steady-state flows and stocks of the four metals

An inventory was made of the flows and stocks of the metals in 1990. These data were entered in FLUX and the FLUX balancing procedure was used to generate a complete overview conforming in terms of mass balance. The complete FLUX overview for zinc in the Netherlands in 1990 is presented in Annex 1 as an example. Similar overviews are available for the other metals. In Annexes 2, 3 and 4 a summary list of the flows and stocks of copper, cadmium and lead has been added. As can be seen from Annex 1, such an overview is quite extensive and difficult to interpret. This is the reason for using indicators, which are derived from the overview. In this section, we present a summary and a brief characterisation of the main results.

Figure III.1.1 shows a substance flow diagram for the four metals.

From this overview a number of conclusions can be drawn:

- The flows of copper, zinc and lead are several magnitudes larger than the flows of cadmium.
- For all metals, we can view the Dutch economy as an open economy: the imports and exports are large compared to the domestic flows, especially in production and refinery. This makes it difficult to connect the use and waste flows to the production and refinery activities within the Netherlands in an analysis. Especially for cadmium this is striking: the rather large amount extracted from the zinc ore is exported in its entirety, while virtually all cadmium-containing products entering the use phase are imported.
- For all metals, we may observe that the emissions are indeed small compared to the flows through the economy.
- For all metals, we can detect a relatively large accumulation in the economic system.
- Although there are no mines in the Netherlands, major flows can be found in the 'extraction' phase for zinc and cadmium. These are the flows connected with the major zinc refinery located in the Netherlands, where cadmium is extracted from zinc ore as well as zinc.
- For copper, zinc and lead, a significant 'back-flow' can be observed. This indicates that a large fraction of the waste materials is recycled. For cadmium, this is not the case: waste materials either end up in landfill (accumulation in the waste treatment phase) or are re-used as building and road construction materials containing cadmium as a trace contaminant.

If we take a closer look at the underlying figures, such as can be found in the Annexes 1 – 4, it can be concluded that functional 'bulk' applications are dominant for copper, zinc and lead, while for cadmium the 'trace' applications are more important.

Figure III.1.1 Substance Flow Diagrams for copper, zinc, lead and cadmium in the Netherlands, 1990.

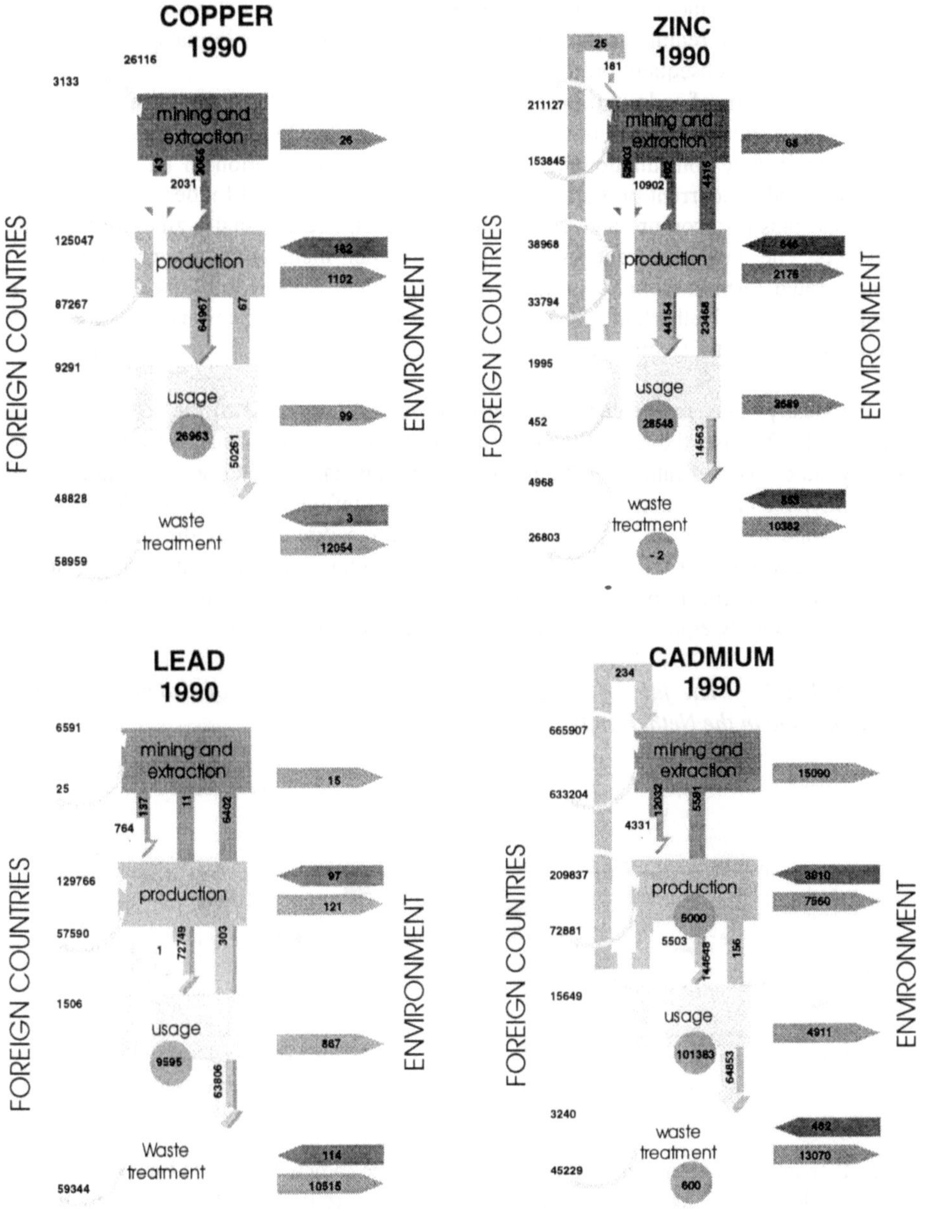

Copper, zinc and lead have major applications in the built environment: gutters, roofs, fences, wiring, pipes, etc. Specific applications include lead in batteries, copper in public transport overhead wiring and zinc in brass products. 'Trace' applications can also be found: lead in petrol, copper in ships' anti-fouling paints, zinc in textiles and tyres, copper and zinc as fodder additives. Comparatively minor flows can be found in the non-functional realm: all three metals are trace contaminants in phosphate rock and iron ore and the subsequent products.
Main applications of cadmium, on the other hand, belong to the 'trace' category: use as a pigment or stabiliser, or in surface coating, for example. Quite a significant part of Cd flows are in fact non-functional, i.e. represent Cd contamination in all sorts of raw materials and concurrent products: phosphate rock, fertiliser and fodder additive; fossil fuels and plastics; iron ore and iron & steel products; and especially zinc ore and zinc products. The only important product group in which metallic cadmium is applied in concentrated form is nickel-cadmium batteries, which in 1990 was a rapidly growing application in the Netherlands, and which is one of the few applications still allowed by the Dutch Cadmium Decree.

In the steady state, major changes may be observed compared to the 1990 situation. Figure III.1.2 shows the fate of the total net inflow of the four metals in 1990 and in the steady state, divided into three categories: emissions, landfill and accumulation in the economy. We can see that compared to 1990 the net inflow is significantly lower in the steady state for three of the four metals. Only for cadmium does the inflow remain more or less constant. However, both landfill and emissions rise, since in the steady state there is obviously no longer accumulation. For zinc we see that in the steady state emissions roughly equal landfill. An evaluation is presented below.

Figure III.1.2 The fate of the 1990 and steady-state inflow of cadmium, copper, lead and zinc in the Netherlands.

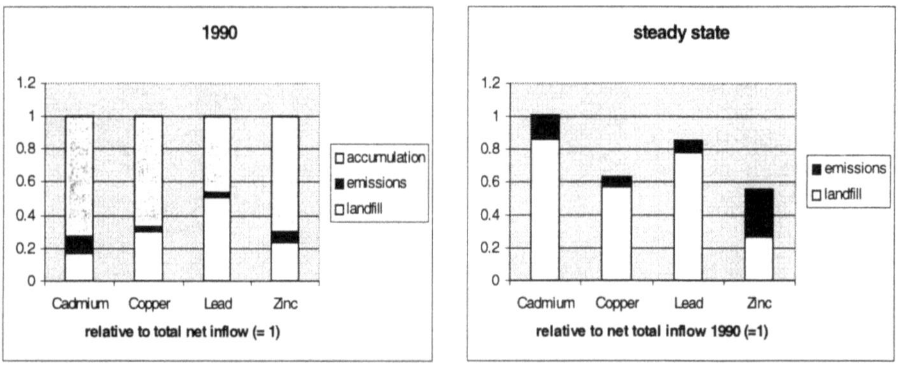

A further interpretation can be made by applying the indicators defined in Section II.6. The results of this exercise are presented in the next three sections, related to the research questions of Part I.

III.1.3 Fate of the mined metals

The indicators developed for the fate of the metals, as described in Section II.6, are:

- total emissions
- total landfill
- accumulation in the economy
- pollution export

All these indicators are applied here.

Total emissions

From Figure III.1.2 it can be concluded that in most cases emissions constitute only a small part of the total fate of the metals. This does not imply that these emissions cause no problems: metals are toxic even in small doses. An evaluation of the harmful potential of these emissions is given in the next section: III.1.4.

As Figure III.1.3 below demonstrates, the emission indicators show an increase of emissions in the steady-state situation compared to the 1990 situation, which is marked especially for zinc. A further break-down of the indicators has been performed for the different environmental media.

Figure III.1.3 Emissions of heavy metals in the Netherlands, 1990 and steady state.

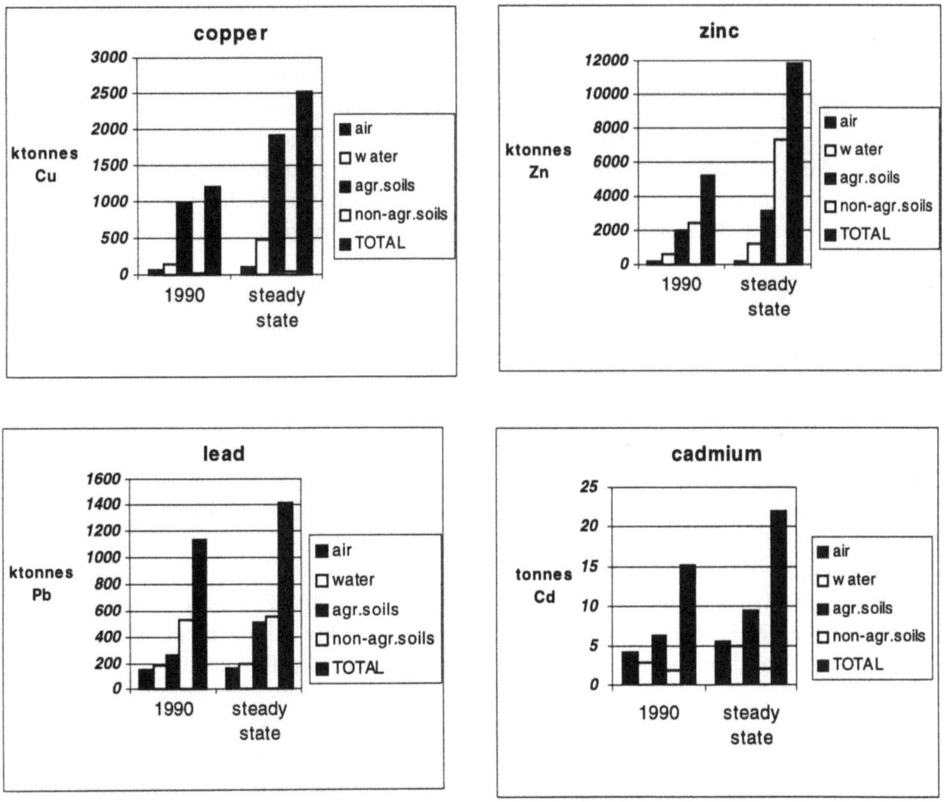

It appears that the increase of air emissions in the steady-state situation compared to the 1990 situation is generally moderate. For cadmium the increase is due mainly to the incineration of spent NiCad batteries, for copper to overhead railway wires. For zinc, air emissions are lower in the steady state, since the amount of zinc in galvanised iron is decreasing. For all four metals, the increase of water emissions in the steady-state situation compared to the 1990 situation is due mainly to the corrosion of metals in building materials (e.g. zinc gutters, galvanised steel, tap water heating equipment and bulk materials such as concrete). However, with respect to the total input to water, it is not emissions within the Netherlands but the inflow of metals from outside the Netherlands via rivers like the Rhine and Meuse that constitutes the dominant source for all four metals (up to over 70%).

The increase of steady-state emissions to agricultural soils compared to 1990 emissions is significant for all metals and is due to increasing flows of organic manure and of source-separated vegetable, fruit and garden waste (the latter being less relevant for lead). The ultimate source behind these increasing flows of copper and zinc is animal fodder. It appears that in the steady-state situation the agricultural soil emissions of copper and zinc are due overwhelmingly (about 80-90%) to the addition to fodder of copper and zinc, respectively. This is an example of closed-loop accumulation (CLA): copper and zinc are added to fodder, which is imported from abroad and fed to Dutch cattle. The manure produced by the cattle, including its copper and zinc content, is spread on agricultural land as an organic fertiliser. Soil concentrations of copper and zinc consequently rise and, with them, copper and zinc levels in maize, pit grass, fresh grass and hay. The livestock are additionally fed with maize, pit grass, fresh grass and hay, and the metals are thus returned to the economy. The eventual steady-state soil concentration due to this cycling of copper and zinc leads to several risk ratios above 1. For non-agricultural soils, a large increase is apparent for zinc only. This is due mainly to corrosion of building materials, which is expected to increase. Apart from that, losses from landfill sites are expected to rise (see below), as well as emissions from applications of waste materials in building and road construction.

It should be noted that all steady-state indicators presented in this section are based on 1990 data and do not take into account any effects of policy measures taken since. For example, the decrease of lead in fuel and the decrease of cadmium in zinc gutters have not been taken into account in the current steady-state results. The use of copper and zinc in fodder has also been reduced since 1990, but this has been neutralised by an increase of the Dutch pig stock, resulting in a higher flow of copper and an equal flow of zinc in fodder in 1994 (Westhoek et al., 1997) compared with 1990. The closed-loop-accumulation example of copper and zinc in fodder is thus still valid.

Total landfill

From Figure III.1.2 it can be seen that the amount of landfilled metals is significantly higher in the steady state. At present, landfill equals roughly a third of net inflow, while it will ultimately almost equal the total inflow. Whether this landfill in itself is considered problematical is another issue. Increased landfill may lead to increased leaching to the environment. In the steady state, because of the balance requirement, the amount of leaching necessarily equals the total amount of landfilled metals. This is not included in the indicators, however, since the time scale on which something of this nature might happen is not relevant for environmental decisions. When introduced in Dynabox, such leaching is found to make only a minor contribution to the health risk indicators from the second group (see Section III.1.4), even in the steady state. The squandering of scarce resources may also be regarded as a problem. There are those who consider landfill sites as mines for the future (Brunner, 1999). Others see this in a less favourable light. Last but not least, when we assume a steady demand, we must also assume that every landfilled kilogram must be replaced by newly produced metals, which implies generation of the emissions associated with production processes.

Accumulation in the economy

For all four metals, accumulation in the economy is at present the most important sink. Relative to gross inflow, the accumulation in the economy ranges between 7 and 14%, being highest for copper and lowest for lead. Relative to net inflow, as can be concluded from Figure III.1.2, accumulation ranges from roughly half to three-quarters. In the steady state this accumulation will have disappeared and will have resulted in increased landfill and emissions. An indication of the time it will take to reach the steady-state situation in the economy can be obtained from the life spans of the products and applications involved. For example, the average life span of functional applications such as building materials lies somewhere between 30 and 50 years, while for non-functional flows of metals in bulk building materials such as concrete this may be a century.

Pollution export

The results for the pollution export indicator are presented in Figure III.1.4. To be able to compare the results for the different metals, the 'footprint' pollution was used as a reference. Actual pollution, i.e. emissions plus landfill, is greater than the footprint in every case. Pollution export is therefore negative. This indicates that the Netherlands is, and will remain, a net importer of pollution for cadmium, copper and zinc. This means there is no net shifting of problems to other countries. For lead this indicator has not been calculated, since it only gives useful information if the economic processes in the region are more or less representative of average economic processes in the world, which is not the case for lead extraction and refining processes in the Netherlands.

Figure III.1.4 Pollution export indicator for cadmium, copper and zinc in the Netherlands, 1990 and steady state.

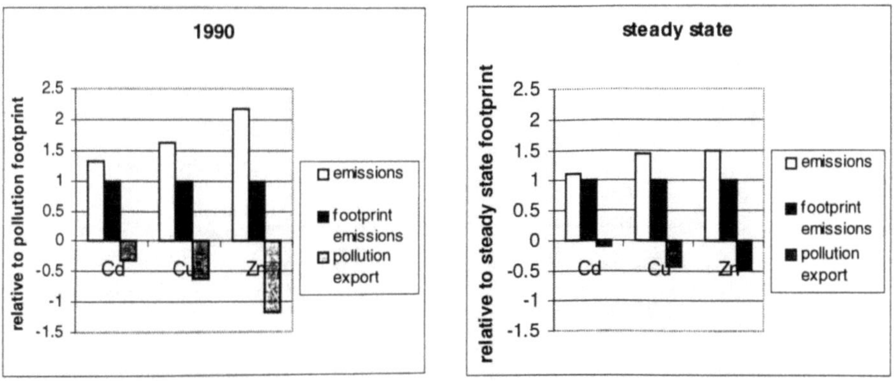

III.1.4 Evaluation of present management in terms of sustainability

The indicators for the evaluation in terms of sustainability are the following:

- environmental concentrations *(PEC/PNEC)*
- human intake *(PDI/TDI)*
- environmental accumulation.

Environmental concentrations(PEC/PNEC)

Figure III.1.5 shows the risk ratios for aquatic ecotoxicity, Figure III.1.6 for terrestrial ecotoxicity.

Figure III.1.5 Aquatic ecotoxicity risk ratios for cadmium, copper, lead and zinc in the Netherlands, 1990 and steady state.

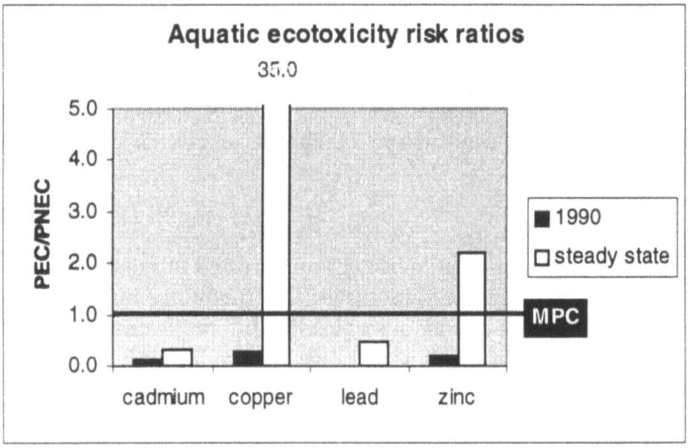

Figure III.1.6 Terrestrial ecotoxicity risk ratios for cadmium, copper, lead and zinc in the Netherlands, 1990 and steady state

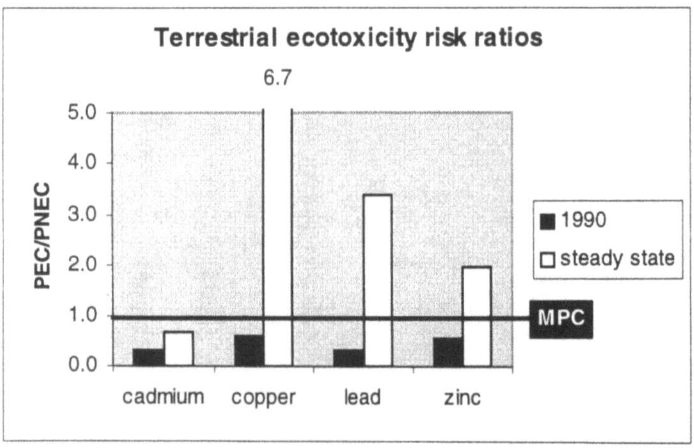

The Dutch Maximum Permissible Concentration (MPC) values have been applied in calculating the risk ratio for aquatic and terrestrial ecotoxicity. The MPC is an ecotoxicological value.[7] MPC values and background concentrations have been taken from Crommentuijn et al. (1997) and Van Drecht et al. (1996). All standards are used as risk indicators; it has not been analysed what exposure levels and effects are actually found at the concentrations calculated. At present, MPC values are not transgressed for any of the metals. However, in the steady state ecotoxicological risk ratios are expected to be over 1 for all metals except cadmium.

Human intake: ADI/TDI

Figure III.1.7 shows the human health risk indicator: *PDI/TDI.*

Figure III.1.7 Human toxicity risk ratios for cadmium, copper, lead and zinc in the Netherlands, 1990 and steady state.

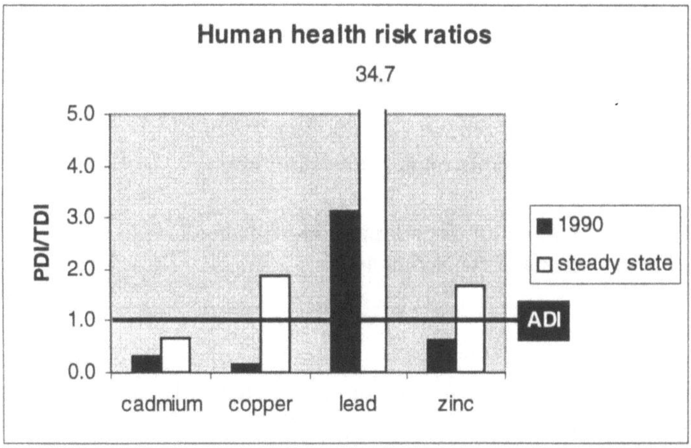

Acceptable Daily Intake (ADI) values defined by the WHO and Tolerable Daily Intake (TDI) values similarly defined by Vermeire et al. (1991) and Cleven et al. (1992) have been applied in calculating the risk ratio for human toxicity.

In 1990, only the ADI for lead is transgressed. This risk may be expected to decline or disappear entirely since it is due principally to lead in petrol, which is being phased out altogether. However, in the steady state the ADI is transgressed for three of the four metals, which indicates that the present metals management regime will lead to health risks in the long run.

Environmental accumulation

About 50% of the environmental inflow of copper and zinc and about 25 % of the inflow of cadmium and lead accumulated in the environment in 1990. This means that

---

7 In the Netherlands a discussion is progress on the derivation of ecotoxicological values for zinc that also take into account the essential significance of zinc for human and other life. The discussion may result in new ecotoxicological values, but these have not yet been proposed (Gezondheidsraad, 1998).

the environmental stock is growing rapidly, which explains the high risk ratios in the steady state. The speed of accumulation can be used to comment on the transition period, i.e. the time it takes to reach a risk ratio of 1. The transition periods for the various metals are also shown in Table III.1.1 below. In calculating the transition periods, current background levels in the various environmental media have been taken into due account. The transition periods vary from 0 years for lead to reach the ADI, to 1000 years also for lead to reach the aquatic MPC level. The transition times for copper and zinc in aquatic ecosystems appear rather short. The results for soil have been compared with the results of the more sophisticated D(SC)B model (see Section III.3) and appear to be fairly similar.

Table III.1.1 Transition period for risk ratios for cadmium, copper, lead and zinc in the Netherlands (years).

	cadmium	copper	lead	zinc
MPC aquatic	∞	3	1000	16
MPC terrestrial	∞	30	550	120
ADI	∞	460	0	130

III.1.5 Design of a sustainable management regime

Five indicators were defined for this purpose. As explained in II.6, only two are applied to the case of heavy metals in the Netherlands:

- technical efficiency
- recycling rate.

For technical efficiency, the results are shown in Figure III.1.8 below, for the basic year 1990 and the steady state. The efficiency of the extraction and production stages is generally high. This indicates that, in order to prevent emissions, not much can be gained by a further boost of industrial efficiency. Comparing the steady-state efficiencies to the 1990 efficiencies, we see a decrease in the use and waste-management efficiencies - due, for example, to corrosion of asphalt, cement and concrete in (utility) buildings, overhead rail wires, cement and landfill emissions - for all metals.

Figure III.1.8 Technical efficiency of the life-cycle stages of cadmium, copper, lead and zinc in the Netherlands, 1990 and steady state.

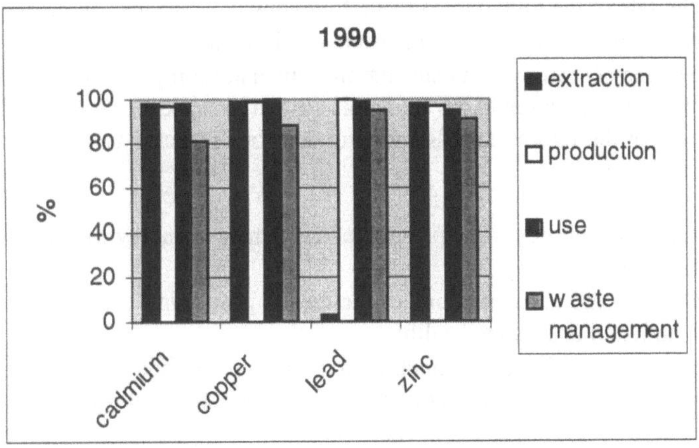

Figure III.1.9 shows the results for the recycling rate. For copper, lead and zinc, functional recycling is at present quite high. There is no significant change in the steady state. The recycling rate is determined largely by the recycling of building materials. A boost of recycling seems difficult, since only the dissipative applications are not recycled at present. For cadmium, recycling refers mainly to various types of NiCad batteries. Because of the present rapid growth of the stock this recycling is expected to increase. Non-functional recycling, i.e. in reused waste materials such as fly-ash, slag and compost, is quite high for cadmium. This might lead to problems in the future.

Figure III.1.9 Recycling rate for cadmium, copper, lead and zinc in the Netherlands, 1990 and steady state.

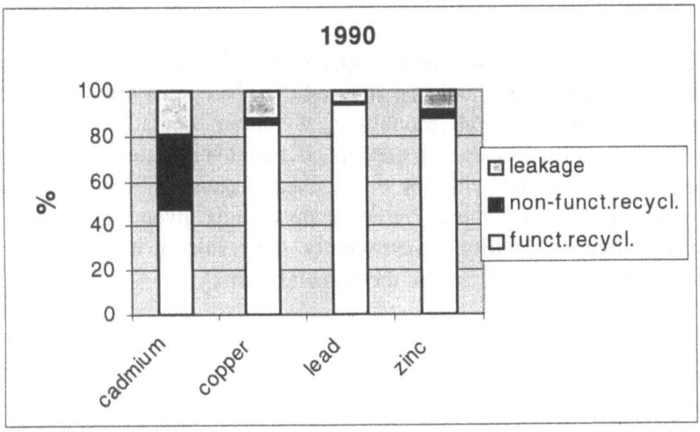

III.1.6 Conclusions

Regarding the fate of the mined metals, the following conclusions can be drawn:
- the Netherlands does not engage in problem-shifting to other countries
- the fate of the net inflow of metals into the Dutch economy is, in decreasing order of magnitude: accumulation in the economic system in stocks of products; landfill; and emissions. In the future, landfill and emissions are expected to rise significantly as the stocks increase.

Regarding the sustainability of the present metals management regime in terms of risks, we conclude the following:
- the 1990 flows and accumulations of the metals pose significant long-term risks to human health and ecosystem health
- for all metals, the built environment, agriculture and landfills are the most important sources of the increase in emissions for the steady-state situation based on the 1990 regime.

In contrast to the apparent general view that these metal flows are well under control, the conclusions of this case study point in a different direction. The problem is all the more pressing since the recycling rates of the metals are already quite high.

From the indicators for the design of a sustainable management we conclude that:
- the increase in emissions takes place despite quite high efficiencies and substantial functional recycling rates; apart from Cd, these were at least 80% in 1990
- the non-functional recycling flows are a major cause of diffuse emissions to the important media (air, water, agricultural soil) for human and ecotoxicity.

A further increase of efficiency and functional recycling is therefore difficult to achieve, and might not be effective, moreover. In fact, relatively small concentrations in specific flows in the economy cause a marked increase of risks through a closed-loop accumulation process, as in the example of Cu and Zn in fodder. A management strategy must therefore look in other directions. This will be discussed further in Part IV.

Although the models used include the full spectrum of flows and accumulations in the economy and environment, it should be stressed that the results are merely indicative. Besides the uncertainties in the modelling, a further limitation is that resource availability has not yet been taken into account. In fact this is assumed to be infinite. So perhaps the high risk ratios will not be reached because of enforced declines in resource extraction. However, this is by no means certain, given the continually rising estimates of resource availability. Consequently, the results at least imply a warning signal as to the sustainability of current metal metabolism.

References

- Annema, J.A., E.M. Paardekooper, H. Booij, L.C.F.M. van Oers, E. van der Voet & P.A.A. Mulder (1995). Stofstroomanalyse van zes zware metalen - Gevolgen van autonome ontwikkelingen en maatregelen. RIVM report no. 601014010, Bilthoven, the Netherlands.

- Brunner, P.H. (1999). Editorial: In Search of the Final Sink. Envir. Sci. & Pollut. Res. 6 (1) p.1.
- Cleven, R.F.M.J., J.A. Janus, J.A. Annome on W. Slooff (1992). Basisdocument zink. RIVM report no. 710401019, Bilthoven, the Netherlands
- Crommentuijn, T., M.D. Polder & E.J. van de Plassche (1997). Maximum Permissible Concentrations and metals, taking background concentrations into account. RIVM report no. 601501001, Bilthoven, the Netherlands.
- Drecht, G. van, L.J.M. Boumans, D. Fraters, H.F.R. Reijnders & W. van Duijvenbooden (1996). Landelijke beelden van de diffuse metaalbelasting van de bodem en de metaalgehalten in de bovengrond, alsmede de relatie tussen gehalten en belasting. RIVM report no. 714801006, Bilthoven, the Netherlands.
- Gezondheidsraad (Commissie Risico-evaluatie van stoffen) (1998). Zink. Gezondheidsraad publicatie no. 1997/34, Rijswijk, the Netherlands.
- Guinée, J.B., J.C.J.M. van den Bergh, J. Boelens, P.J. Fraanje, G. Huppes, P.P.A.A.JH. Kandelaars, Th.M. Lexmond, S.W. Moolenaar, A.A. Olsthoorn, H.A. Udo de Haes, E. Verkuijlen & E. van der Voet (1998). Evaluation of metal flows and accumulation in economy and environment. Ecological Economics, in press.
- Vermeire, T.G., M.E. van Apeldoorn, J.C. de Fouw en P.J.C.M. Janssen (1991). Voorstel voor de humaan-toxicologische onderbouwing van C-(toetsings)waarden. RIVM report no. 725201005, Bilthoven, the Netherlands.
- Westhoek, H.J., L. Beijer, W.J. Bruins, P.H. Hotsma, J.W.M. Janssen & E.J.R. Maathuis (1997). Aan- en afvoerbalansen van zware metalen van Nederlandse landbouwgronden. Informatie- en KennisCentrum Landbouw, Ede, the Netherlands.

III.2 Applications of Material-Product Chain analysis

Patricia Kandelaars & Jeroen van den Bergh

III.2.1 Introduction

The diversity of potential approaches and applications of M-P chain analysis is illustrated here by discussing five specific applications (for a more comprehensive overview, see Kandelaars 1998). Each of the applied models combines elements of both physical flow and economic models in a particular way. It is important to know what model type to use for which purpose. The main focus of the applications A to D is the demand for a service that may be met with different products (partial), while in application E the whole economy, including production sectors and household groups is allowed to change (general). Table III.2.1 shows which type of models have been used in the applications discussed in this section, as well as which "pure model types" of Table III.2.1 they integrate.

Table III.2.1 An overview of the physical and economic models that are used for the application of M-P chain analysis.

	Static optimisation model (Appl. A)	General equilibrium model (Appl. B)	Dynamic simulation model (Appl. C)	Dynamic simulation model (Appl. D)	Applied general equilibrium model (Appl. E)
MFA/Physical I-O analysis	+	+	+		+
LCA				+	
Economic models of natural resources			+	+	
Pollution models	+	+			+
Economic I-O models					+
Macroeconomic models					+
Technological change and evolutionary models	+				

Each of these model categories is described in the subsequent sections.

III.2.2 Application A: A static optimisation model of rain gutters

The first application is a static optimisation model of an M-P chain in which an `environmental manager' optimises the costs under a set of physical and technological

restrictions (Kandelaars and Van den Bergh, 1996). This model is of the type as described in section II.3. The model optimises the costs for the demand of a service which includes, besides recycling and reuse, demand and production functions. The goal of this application was to explore how policies or strategies that are applied to different stages of an M-P chain may differ in their impact on a number of physical and economic indicators. The model includes MB conditions for each activity and process. The material balance conditions are important to determine the amount and the type of waste material that is generated after a product is disposed of.
The environmental manager is supposed to minimise the total costs of satisfying the demand in an M-P chain. This model is a combination of a material flow analysis (MFA) (or physical I-O analysis) and elements of a pollution model (environmental manager, waste/pollution). The "total costs" are defined as the costs of new products, reused products and waste dumping. It is possible to include external costs as well. In this model, the production process has as inputs two types of materials that may be new or recycled, and as outputs new products and waste material. The waste material from the production process may be recycled together with the waste material resulting from the disposal of the product. The choices that need to be made concern: which and how much new and recycled materials of each type in the production process; how many new and reused products for meeting the demand for the service; and, which percentages of the materials to be recycled. MB conditions are included at the level of the product, the production process and the system. The MB conditions are important to keep track of the material content of products and the waste material that is generated after a product is disposed of. The model may include endogenous prices and different production technologies. With this type of model the optimal recycling and reuse rate, and the optimal input mix of materials in the production function may be assessed. Furthermore, the impact of policy instruments (including physical restrictions, such as the minimal rate of recycling) on material dumping, recycling, reuse of products and the costs of meeting the demand can be analysed.

The model has been applied to an M-P chain analysis of rain gutters. The problem analysed is how the use of zinc in zinc gutters may be reduced, because the use of zinc gives environmental problems (Gorter, 1994). A distinction was made between zinc and PVC gutters. The basic optimisation model is linear. The production functions are linear, with one input resulting in a fixed amount of waste material for each type of gutter. All waste material is recycled as long as the recycled material is cheaper than the new material, otherwise no recycling takes place. These extremes set the percentages of materials that are recycled. In a linear model the optimal allocation of the demand over the two types of rain gutters is either only zinc or only PVC gutters. Alternative policies, such as a product charge, recycling standards and subsidies, may change the distribution of demand resulting in a change in the extraction of materials and the total costs of the M-P chain. In a non-linear model the demand is met by a combination of zinc and PVC gutters. In the scenario in which the price of waste treatment depends on the number of zinc gutters (which makes the model non-linear), the demand is met by both zinc and PVC gutters. In this scenario, the levy on waste treatment (i.e. a levy on the dumping of galvanised steel) increases the net costs and a certain proportion of the zinc gutters will be substituted by PVC gutters. The results show that the current situation is not optimal, in the sense that the net costs of the demand are not at their minimal level.

Table III.2.2 Performance of indicators under different scenarios.

Scenarios / Indicators	Base	Product charge	Recycling	Product charge collection subsidy	Waste + treatment	Current situation
Net costs (millions of Dfl)	19.8	23.2	19.6	21.8	20.1	20.5
Zinc gutters (units)	129,000	0	129,000	0	75,600	103,200
PVC gutters (units)	0	129,000	0	129,000	53,400	25,800

The application shows that for an analysis of policies related to materials or products, the demand for products and materials needs to be considered simultaneously. A policy imposed on a product or material may affect the M-P chain at different, but connected levels: for instance, the material, product and demand level. The model minimises the total costs of the system for different (policy) scenarios, and it may be used to analyse the differences in total costs, extraction and waste disposal of each (policy) option.

III.2.3 Application B: A general equilibrium model with materials flows and environmental externalities

A second application is based on a theoretical general equilibrium analysis of an M-P chain. The general equilibrium model is not an economy-wide model, but a model for a specific M-P chain (Kandelaars and Van den Bergh, 1997a). The application integrates elements of MFA and environmental policy analysis. In other words, a welfare-economic perspective is integrated with a physical perspective. The model describes extraction, production, recycling, consumption, waste treatment activities and MB conditions in a general equilibrium framework. Although some studies (Fullerton and Kinnaman, 1995; Lusky, 1975 and 1976; Sullivan, 1987; Dinan, 1993; Morris and Holthausen, 1994) have mixed some of these elements, the combination, as pursued in this application, is new. In the market equilibrium these externalities are optimised by imposing adequate taxes.

The economic activities are represented in the equilibrium model via separate profit and utility maximisation formulations, and technical, budget and material balance restrictions. With this structure three important questions may be addressed: Via which price correction instruments or combination of such instruments (policy packages) can the external costs associated with resource extraction and waste disposal be optimised? What are the optimal price correction rules supporting the alternative policy packages? And, which second-best rules apply when specific instruments are not available or cannot be used?

The M-P chain has the following detailed structure. A main product is generated by a production process that uses three inputs: capital, new material and recycled material. The new material is obtained through the extraction of natural resources using capital. The alternative product only uses capital as an input. The two products are bought by n identical households. After use of the products in the consumption process these end up as garbage or as waste material entering a recycling process. This recycling process uses capital to recycle waste material, and generates waste material itself. The latter is treated in a waste treatment process together with the garbage directly originating from consumption. This leads to the distinction between harmful and non-harmful waste. The demand for the product by consumers results in two types of externalities: related to extraction and to harmful waste dumping. These externalities have a negative impact on the utility of the households: for example, cutting trees reduces the forest area and may reduce biodiversity, and waste dumping may cause health risks.

The results show that the externalities caused by extraction and generation of harmful waste can be optimised by imposing a tax on new (raw) materials. In a second-best world these externalities may be optimised by imposing taxes on harmful waste and on the use of recycled materials. The optimal taxes on the generation of harmful waste and on new materials depend partly on the same term that includes prices and marginal products of production and waste treatment functions. This implies that a change in some variables causes a shift of taxation from the start (extraction activity) to the end (waste treatment activity) of the chain. This linkage implies that the whole M-P chain needs to be considered when analysing optimal policy packages. Therefore, to derive the optimal rules for taxes and subsidies it is necessary to consider flows and processes related to raw materials, recycled materials, main products, garbage from consumption, and material and recycling waste.

This model includes MB conditions for every economic activity in the M-P chain. In a production function with various inputs, MB conditions are important to keep track of the material content of products. Once a product is disposed of, it needs to be transformed and decomposed into the one or more material flows of which they were originally made. If more types of materials are used, it is relevant to distinguish between various types of waste material. When a production function has several outputs, for example, products and production waste, MB conditions are needed to determine the material content of the products and the amount of production waste. Thus, a multi-input or multi-output function (e.g. a production, recycling or waste treatment function) requires that for each type of material (or product) the input equals the output.

The relevance of including the M-P chain, from extraction to waste treatment and the linkages between all stages, is apparent from the variables in the optimal tax rules. Moreover, the analysis shows that the externalities generated by extraction and waste treatment are connected.

III.2.4 Application C: A dynamic simulation model of rain gutters

In a third application a dynamic descriptive model is presented to study the impacts of economic and policy developments aimed at reducing the use of materials (Kandelaars *et al.*, 1996). The consequent dynamic analysis allows one to address processes such as accumulation and delays of durable products. Accumulation of products and materials may occur in the economy: for example, for durable products there is a time-lag between the purchase and the disposal of the product. Delays may occur between the implementation and the effect of a policy. The model traces the effects of changes in material flows over time: for example, through substitution and recycling. Using the model, policy scenarios were simulated to assess their influence over time on an M-P chain. The model was applied to rain gutters.

Rain gutters are durable products with a lifetime of several years. To distinguish between products that are produced in different years, a vintage approach is adopted to describe rain gutters. The model describes the demand for rain gutters which depends on the number of new houses to be built and the number of houses to be renovated. This demand is allocated to zinc or PVC rain gutters. The model analyses the demand for zinc and PVC gutters and the material flows associated with this demand, under various (policy) scenarios. Simulation results show that a shift in consumer preference from zinc to PVC gutters leads to a reduction in the use and disposal of zinc, but to an increase in the disposal of PVC (see Table III.2.3).

Table III.2.3 Overview of the results of the scenarios.

Scenarios Indicators		Preferences shift	Product charge	Recycling	Mixed
Economic	Demand for gutters	+	+	0	+
	Allocation variable	-	-	0	-
	Net costs	+	+	-	+
Environmental	Recycling zinc (%)	0	0	+	+
	Recycling PVC (%)	0	0	+	+
	Effect on stock zinc ore	++	++	+	++
	Zinc waste in environment	-	-	-	-
	PVC waste in environment	+	+	-	+
	Galv. steel waste in envir.	≈	≈	≈	≈

However, over time the demand for gutters increases because the lifetime of PVC gutters is shorter than that of zinc gutters. The product charge scenario imposes a charge on zinc gutters. Subsidies on recycled zinc and PVC do not affect the demand or the allocation of gutters, but have a positive effect on the extraction of zinc ore and the disposal of zinc and PVC waste into the environment. This dynamic descriptive model for gutters reflects the importance of products and processes for the analysis of flows of materials, tracks the impact of economic and government policy variables on M-P chains, and includes dynamic aspects such as accumulation and delays.

III.2.5 Application D: A dynamic simulation model of window frames

In the fourth application a dynamic analysis of M-P chains was performed in which a life-cycle assessment (LCA) was combined with an economic analysis. The dynamic descriptive model includes the environmental impacts of the M-P chain (Kandelaars and Van den Bergh, 1997b). Also here accumulation and delayed effects are considered. The dynamic model was applied to window frames. The demand for window frames is allocated to four types of window frames. The distribution over the various types depends on whether the window frames are for newly built houses or for renovated houses. Window frames are durable products with a long lifetime, so that also here a vintage approach was adopted. Three economic agents are modelled, namely producers (of window frames), consumers (e.g. a construction firm) and a regulator (e.g. a government).

In the model two policy packages are imposed aiming to reduce (1) the depletion of raw materials, and (2) water pollution. The policy package for reducing the depletion of raw materials consists of levies that are imposed on hardwood window frames and on the dumping of hardwood, and information provided to consumers to persuade them to choose pinewood window frames. To reduce water pollution, the policy package consists of levies on the use of new aluminium and PVC, on the dumping of aluminium and PVC, and on the use of aluminium window frames. As in (1) above, information in favour of pinewood window frames is provided to consumers.

Table III.2.4 shows that imposing a mix of policies on depletion may reduce the raw materials depletion by more than 50% in 20 years compared with the scenario without policies. Also other environmental aspects are positively affected. The policy package on water pollution has a delayed effect, because water pollution mainly occurs at the waste treatment stage of the M-P chain. Table III.2.5 shows that after 40 years, water pollution is reduced by 30% compared with the base scenario. The package also reduces other environmental impacts substantially compared with a scenario without policies. Both policy packages have a positive effect on the average cost per window frame paid by the consumers. The revenues for the government from the levies are positive but decreasing over time, because consumers buy less of the product on which a levy is imposed. A multi-criteria analysis was performed to see the effects of single instruments on the environmental indicators. This analysis shows that with equal weights for each indicator, a levy on aluminium window frames is the best single instrument. This type of model can be used to analyse the effects of a specific policy (package), aimed at reducing a certain environmental indicator, on other environmental and economic indicators. These total effects need to be taken into account simultaneously.

Table III.2.4 The influence of policy packages on the depletion of raw materials
(in 10^6 depletion units)

Year	Base case	Package aimed at resource conservation	Package aimed at water pollution reduction
1990	6051	6051	6051
1995	4979 (100%)	3897 (78.2%)	4306 (86.5%)
2000	3256 (100%)	2009 (61.7%)	2192 (67.3%)
2010	3279 (100%)	1607 (49.0%)	1695 (51.7%)

Table III.2.5 The influence of policy packages on water pollution.
(in 10^6 m^3 potentially polluted water)

Year	Base case	Package aimed at resource conservation	Package aimed at water pollution reduction
1990	483.4	483.4	483.4
2010	483.2 (100%)	469.8 (97.2%)	395.8 (81.9%)
2030	397.5 (100%)	359.7 (90.4%)	255.2 (64.2%)
2050	399.0 (100%)	367.7 (92.2%)	259.4 (63.4%)

Under the resource conservation scenario a levy is put on the dumping of hardwood and on the use of hardwood window frames. Because of increased information about window frames the consumers are assumed to change their preferences over time. Hence, the distribution of the demand over the four types of window frames also changes over time, as can be seen in Figures III.2.1 and III.2.2 for new and renovated houses, respectively. The preferences change smoothly over time.

Figure III.2.1 The allocation of the demand for window frames for renovated houses in Scenario 2.

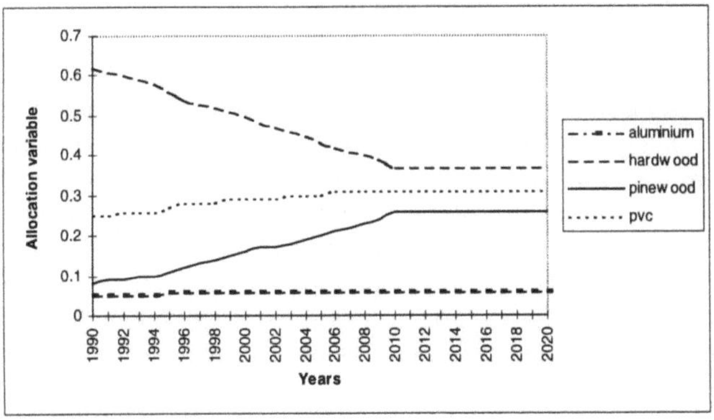

Figure III.2.2 The allocation of the demand for window frames for new houses in Scenario 2.

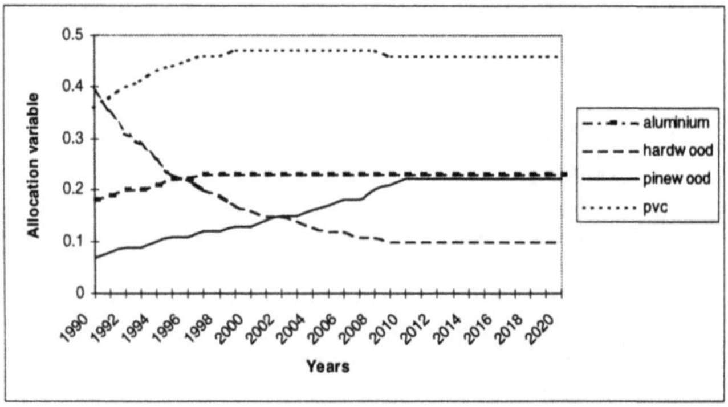

The total demand is lower than in the reference scenario because the levy on the use of hardwood window frames increases the average price of window frames, so that a certain proportion of the houses are subjected to renovation of window frames.

III.2.6 Application E: Linking FLUX and an AGE model for zinc and lead

The fifth application presents a first "empirical step" in filling the gap between physical and economic models, by combining a material flow model for zinc and lead with an applied general equilibrium (AGE) model that are both calibrated for the Netherlands. The goal of this model was to empirically assess the economy-wide and environmental effects of environmental policies focusing on the environment (Kandelaars and Dellink, 1998). The AGE model used for the model is the "Taxinc-model" which is a static AGE model with 61 production sectors and 44 household groups (Keller, 1980; Cornielje, 1990). It allows the analysis of the sectoral and distributional consequences of policies. The material flow model to describe zinc and lead is FLUX (see Section II.2).

To combine the Taxinc-model and FLUX, the economic sectors of FLUX are connected to those of the Taxinc-model. A material levy on a sector in the Taxinc-model depends on the use of materials (in kilograms, derived from FLUX) and on their input or output (in monetary units, derived from the Taxinc-model). Thus, the levy combines the physical and monetary units. This approach facilitates the analysis of the effects of material and product policies on various production sectors and household groups, and on trade and employment. The changes in the use of materials are determined on the basis of these effects. In an AGE model, substitution between different production sectors is considered, but, for studying material flows, substitution within a production sector may also be important.

Material policies are imposed to reduce the use of specific materials. Such policies generate revenues, which in this model are redistributed to the tax payers by lowering the labour tax. A material levy and a labour tax reduction shift the tax burden from labour to the use of materials. This shift is attractive because it taxes more what we want less of (use of materials), and less what we want more of (labour).

The policies are formulated for the metals zinc and lead. Table III.2.6 shows that the share of the basic metal industry in the total inflow of zinc is enormous (more than 80% of the total zinc use). Other large users of zinc are the basic chemical industry, metal products manufacturing and trade (through imports). The lead intensity of the basic metal industry is much lower. Apart from these production sectors, the construction sector has a high lead intensity.

At first glance, the large zinc and lead intensity of the other sectors may seem surprising. However, these sectors encompass the waste treatment and processing firms that account for a large use of metals (Boelens and Olsthoorn, 1998). It should be noted that a high (low) metal content of the inputs of a production sector does not necessarily imply a high (low) metal intensity of the goods and services produced in the sector. The reported intensities are based on the place where the metals enter the economic process. For example, a final stage production sector may produce goods and services with a high metal content, but add a little more metal in the production process itself (small inflow). The materials included in the economic inputs are already accounted for in previous production stages.

Table III.2.6 Metal intensities of selected production sectors in the Ntherlands, 1990.

Production sector	Zinc intensity (kg/guilder)	Share of zinc use (%)	Lead intensity (kg/guilder)	Share of lead use (%)
Agriculture	0.01	0.2	0.00	0.0
Grain mills	0.06	0.3	0.00	0.0
Petroleum refineries	0.01	0.1	0.02	0.2
Basic chemical industry	0.44	6.3	0.31	6.5
Basic metal industry	12.30	81.4	4.78	46.0
Fabricated metal products	0.42	3.7	1.08	13.8
Electricity supply	0.03	0.1	0.02	0.1
Construction	0.04	0.8	0.24	6.8
Wholesale and retail trade	0.20	5.7	0.60	24.1
Other transport	0.00	0.0	0.03	0.4
Civil government	0.00	0.0	0.00	0.1
Other services	0.44	1.4	0.42	2.0

The results of the policy analysis with the combination of FLUX and AGE models show that the effects of a regulatory material levy are small for household groups and for most production sectors. The impact of this levy on the basic metal industry and some other large metal-using production sectors may be significant (see Table III.2.7). In theory, the combined effect of the material policy and the labour tax reduction may have a positive effect on the environment (i.e. less material use) *and* on employment, which is called a "double dividend". Here, however, no double dividend was found for any policy scenario.

Table III.2.7 Changes in real output in monetary terms of selected production sectors as a result of policy packages.

Production sector	levy on primary zinc use	levy on zinc throughput	levy on zinc products	levy on primary lead use	levy on primary zinc + lead use
Basic chemical industry	-0.09	-1.62	0.46	-0.13	-0.10
Basic metal industry	-10.51	-0.07	-13.15	-7.01	-9.24
Fabricated metal products	-0.56	-3.41	-0.54	-0.95	-0.71
Electricity supply	-0.46	-0.41	-0.55	-0.37	-0.43

From this illustration of material policies in an economic model it may be concluded that the combination of an AGE model and a material flow model can produce an appropriate tool for analysing environmental, sectoral and distributional effects of material policies. Consistent analysis of the interactions of these effects may be important for environmental-economic policy making.

III.2.7 Further research

Several research questions may be addressed that build on the results of the study of M-P chain analysis, in order to improve the models in an analytical sense or to make these models more appropriate as tools in environmental-policy analysis.

Further research on the behaviour of economic agents in their choice for materials, products, recycling, reuse and waste treatment is required to understand the effects of policies aimed at changing this behaviour. The connection between various stages in the M-P chain, e.g. production, consumption and recycling, needs to be studied to analyse the trade-offs between different stages of the M-P chain.

Consumers and producers can choose between different (new and reused) products, and various (new and recycled) materials. These choices may require the inclusion of the notion of imperfectly substitutable products and materials in the models. Substitution between heterogeneous products and materials needs to be further analysed.
Production functions which represent the transformation of materials into products with the use of capital and labour do not generally take MB conditions into account. In the models of

this study these MB conditions are considered explicitly. In most traditional economic models the material contents of products are not measured. However, in an M-P chain the material contents of products are important: for example, to derive the amount of waste material generated after the product is disposed of. When a production function has multiple inputs or multiple outputs, the amount of products does not determine the amount of materials used. Therefore, with multiple inputs or outputs, it is required to keep track of the material input and output in the production function. When a product contains several materials, the input of these materials into the production function and the possible production waste has to be assessed in order to transform the product into the different types of waste material.

Further research on the dynamic and evolutionary aspects of M-P chain modelling may usefully focus on the inclusion of technological changes, new (imperfect substitutable) materials and products. Dynamic effects related to the accumulation of materials and products in the economy, for instance, due to durable products, and delays in the effects of policies, technological change or other changes, need to be studied in detail to avoid unforeseen difficulties.
With the modelling of M-P chains, the effects of policies may be analysed for various economic activities in order to study the trade-offs between an activity, an environmental problem or a geographical area and another activity, problem or area. In this study, the latter spatial trade-offs are not included. Spatial aspects that particularly require analysis are the geographical scale of imposing policies, and trade-offs between spatial levels: for instance, the shift from generating waste in the Netherlands to other countries.

More empirical research is required regarding the interactions between: economic and physical aspects of M-P chains; the behaviour of economic agents regarding the choice of materials, products, recycling, reuse waste treatment; the relationship between economic activities and environmental problems; and, the trade-offs between the economy and the environment. Empirical research may help to fine tune the models to make them more appropriate as decision-support tools for environmental policy analysis.

References

- Boelens, J. and X. Olsthoorn (1998). Software for material flow analysis. In: P. Vellinga, J. Gupta and F. Berkhout (eds.). Substantial Sustainability, Kluwer Academic Publishers, Dordrecht, forthcoming.
- Cornielje, O. (1990). Rationing and Capital Mobility in Applied General Equilibrium Models, VU University Press, Amsterdam.
- Dinan, T.M. (1993). Economic efficiency effects of alternative policies for reducing waste disposal, Journal of Environmental Economics and Management, Vol. 25:242-256.
- Fullerton, D. and T.C. Kinnaman (1995). Garbage, recycling, and illicit burning or dumping, Journal of Environmental Economics and Management, Vol. 29:78-91.

- Gorter, J. (1994). Zinkbalans voor Nederland 1990. Deel 1: de economische stromen (Balance of zinc for the Nether-lands in 1990. Part 1: the economic flows). In: Kwartaal-bericht Milieusta-tistieken CBS, Den Haag, 1994 (in Dutch).
- Kandelaars, P.P.A.A.H. and J.C.J.M. van den Bergh (1996). Materials-product chains: theory and an application to zinc and PVC gutters, Environmental and Resource Economics, Vol. 8:97-118.
- Kandelaars, P.P.A.A.H., J.B. Opschoor and J.C.J.M. van den Bergh (1996). A dynamic simulation model for materials-product chains: An application to gutters. Journal of Environmental Systems, Vol. 24 (4):345-371.
- Kandelaars, P.P.A.A.H. and J.C.J.M. van den Bergh (1997). Dynamic analysis of materials-product chains: An application to window frames, Ecological Economics, Vol. 22:41-61.
- Kandelaars, P.P.A.A.H. and R.B. Dellink (1999). Economic effects of materials policies: combining an applied equilibrium model with physical flows, Journal of Policy Modelling, forthcoming.
- Kandelaars, P.P.A.A.H., M.H. Jansen and A.J.D. Lambert (1996). Survey of methods of material and product flows analysis, Tinbergen Institute discussion paper 96-124/5, Amsterdam.
- Kandelaars, P.P.A.A.H. (1999. Economic Models of Material-Product Chains for Environmental Policy Analysis , Kluwer Academic Publishers, forthcoming.
- Keller, W.J. (1980). Tax Incidence: A General Equilibrium Approach, North-Holland, Amsterdam.
- Lusky, R. (1975). Optimal taxation policies for conservation and recycling, Journal of Economic Theory, Vol. 11:315-328.
- Lusky, R. (1976). A model of recycling and pollution control, Canadian Journal of Economics, Vol. 9:91-101.
- Morris, G.E. and D.M. Holthausen (1994). The economics of household solid waste generation and disposal, Journal of Environmental Econo-mics and Management, Vol. 26:215-234.Sullivan, A.M. (1987). Policy options for toxic disposal: laissez-faire, subsidization, and enforcement, Journal of Environmental Economics and Management, Vol. 14:58-71.

III.3 Applications of dynamic balances in agricultural systems

Simon Moolenaar & Theo Lexmond

III.3.1 Introduction

Two detailed applications of dynamic balance studies in agro-ecosystems are presented in this section in order to show the possibilities and limitations of the balance approaches that are described in Section II.4. The first case-study focuses on heavy-metal flows in arable farming systems. The second case study describes heavy-metal flows at the field and farm level for mixed farming systems. After drawing some conclusions from these applications, recommendations for further research and for policy are made.

III.3.2 Case-study: Arable farming systems

Arable farming systems are characterized by their crop rotation. In such systems, the most important inputs are mineral (N, P, K) fertilizers, animal manure, organic amendments (sewage sludge and compost), and atmospheric deposition. Output flows consist of produce (crops), inputs to the plough layer and losses from the plough layer by leaching to deeper soil layers and soil erosion (Figure III.3.1).

Figure III.3.1. Arable farming system (dotted lines represent system boundaries)

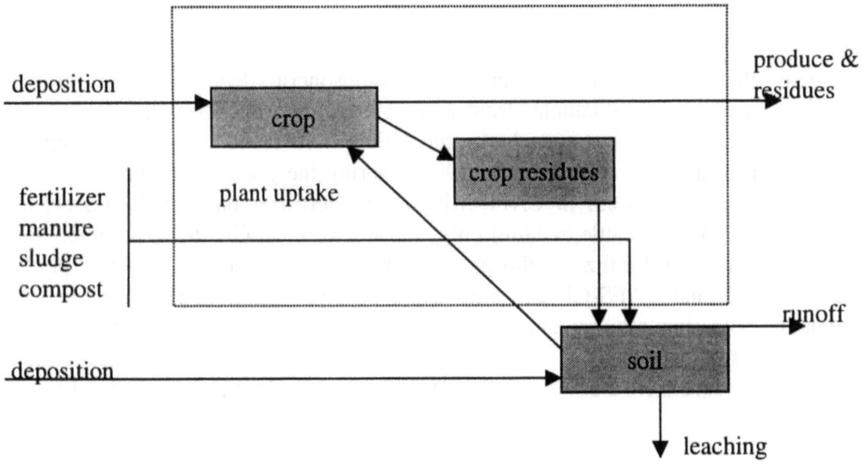

Three arable farming systems as practised at Nagele experimental farm (near Nagele in the Northeast polder; 52°39_ N, 5°44_ E) were chosen for this study i.e., conventional (Conventional Arable Farming System (CAFS): 22.7 ha), integrated (Integrated Arabel Farming System (IAFS): 17 ha), and ecological (Ecological Arable Farming System (EAFS): 22.2 ha). The 4-year crop rotation of the conventional system consists of ware & seed potato, sugar beet, chicory & onion, winter wheat & spring barley. This system comprises two parts, one with mineral fertilizers only (CAFS-MF: 14 ha) and one with both mineral and organic fertilizers (CAFS-OF: 8.7 ha). The crop rotation in the integrated system is similar to that in the conventional system (carrot instead of chicory). The 6-year crop rotation of the ecological system consists of seed potato, spring wheat, celery & onion, spring barley, carrot, and oats (Vereijken, 1992).

Heavy-metal balance sheets for the three systems were calculated. Soil samples were taken from the plough layer (top 30 cm) of each individual field (30 samples in total) and analysed for Cd, Cu, Pb, and Zn. General soil characteristics and average heavy-metal contents that were determined are given in Table III.3.1. Detailed information on soil, fertilizer, and plant analyses is given in Hatziotis (1995) for Cd and Zn and in Van Kuik (1995) for Cu and Pb.

Table III.3.1. General characteristics and heavy-metal contents of the Nagele soil.

pH-KCl:	7.4
Organic matter content (mass %):	2.6
Clay content (mass %):	24
$CaCO_3$ content (mass %):	10
ρ (kg m^{-3}):	1400
Pw –number:	25
K-number:	17
Cd content (mg kg^{-1} dry weight):	0.5 (0.48-0.52)
Cu content (mg kg^{-1} dry weight):	23 (19-27)
Pb content (mg kg^{-1} dry weight):	35 (33-38)
Zn content (mg kg^{-1} dry weight):	100 (99-107)

Inputs

At Nagele, the most important sources are atmospheric deposition and fertilizer application. Irrigation water samples from a nearby ditch did not show detectable levels of heavy metals. Inputs via atmospheric deposition were based on measure-ments at Biddinghuizen near Nagele (Aben et al., 1992). During the growing season 1993-1994, various fertilizers were used: in CAFS-MF mineral fertilizer only, in EAFS organic fertilizers (solid goat and cattle manure) only, and in IAFS and CAFS-OF a combination of organic and mineral fertilizers. Phosphate requirements were met with liquid poultry manure in IAFS and CAFS-OF and with triple superphosphate (TSP) in CAFS-MF. Nitrogen and potassium requirements were met with ammonium nitrate limestone (CAN) and muriate of potash (K-60), respectively. Fertilizer applications and heavy-metal contents in these fertilizers are shown in Table III.3.2 and Table III.3.3, respectively.

Table III.3.2 Manure and fertilizer inputs (kg dryweight) in the growing season '93/'94 compared with surface area (ha) for the farming systems at Nagele experimental farm.

	Ecological	Integrated	Conventional	
			Organic fert.	mineral fert.
Surface area	22.2	17	8.7	14
Poultry manure	0	19695	11263	0
Goat manure	45859	0	0	0
Cattle manure	16352	0	0	0
CAN	0	4250	1782	5367
K-60	0	4887	2322	5211
TSP	0	0	0	2025

CAN: ammonium nitrate limestone; K-60: muriate of potash; TSP: tripel super phosphate

Table III.3.3. Heavy-metal contents in manure and fertilizers (measured numbers: mg kg^{-1} dry weight) used at Nagele experimental farm.

	Cd	Cd (mg kg^{-1} P_2O_5)	Cu	Pb	Zn
Poultry manure	0.42	7.7	72.7	4.9	647
Goat manure	0.38	30.6	45.9	11.9	157
Cattle manure	0.37	20.7	33.2	31.9	167
CAN	0	-	0.45	0.4	1.78
K-60	0	-	0.48	1.1	1.38
TSP	31.4	68.4	43.9	3.9	593

CAN: ammonium nitrate limestone; K-60: muriate of potash; TSP: tripel super phosphate

Crop removal

Total metal removal with crop follows from yield (defined as dry weight at economic maturity stage: bookkeeping data) removed from the fields and its metal content (measurements). Green materials recycled within the farm (such as grass, clover, lucerne, leaves of sugar beet, and leaves of celeriac) are not included in the calculation of total output. Straw was taken off the field and sold. The area-weigthed mean values for crop removal of the 4 farming systems are shown in Table III.3.4. Output in produce is lowest in EAFS since crop yields are lowest. Sugar beet (with a high offtake of Cu, Zn, Cd) is not produced. Furthermore, part of the area is used as so-called ecological infrastructure on which grass/clover/lucerne are grown.

Table III.3.4. Area-weighted mean heavy-metal offtake by crops (g ha^{-1} yr^{-1}) in four arable farming systems at Nagele experimental farm.

	Cd	Cu	Pb	Zn
Ecological	0.6	33.3	1.24	138
Integrated	0.94	48.7	1.58	204
Conventional (organic fert.)	0.78	58.3	2.62	190
Conventional (mineral fert.)	0.82	50.5	2.2	187
Conventional (total)	0.81	53.5	2.35	188

Leaching
Metal solubility is expected to be low in this calcareous soil. As Cd, Cu, Pb and Zn co-exist in the soil solution, we have determined competitive adsorption isotherms for the mixture of Cd, Cu, Pb and Zn at soil pH in order to determine the solute concentrations. The concentration ranges in the mixture correspond to those resulting from deposition and fertilizers. The competitive cations Ca, Na, and K were added in the ratio 3:1:1.

Lead was not present in detectable concentrations (< 16 μg l^{-1}) in the equilibrium solution. Hence, no adsorption isotherm could be constructed. For Cd, Cu and Zn, a Freundlich type relationship between concentration in solution (c) and adsorbed amount (q) could be fitted (Table III.3.5). The initial solution concentrations were calculated by substituting the q_i values (Table III.3.5) in the adsorption models for Cd, Cu and Zn equal 0.02, 2.7, and 0.2 (mg m^{-3}), respectively. Multiplying these concentrations with the precipitation surplus of 0.3 m yr^{-1}, results in leaching rates of 0.06, 8.1, and 0.6 g ha^{-1} yr^{-1} for Cd, Cu and Zn, respectively.

Table III.3.5. Adsorption models for Cd, Cu, and Zn in the Nagele soil expressing the relationship between concentration in solution (c: mg m^{-3}) and the adsorbed amount (q: mg kg^{-1}). The initially adsorbed amount is given by the value of q_i.

	Adsorption model	r^2	q_i
Cd:	$c = 0.19\, q^{1.47}$	0.99	0.24
Cu:	$c = 0.026\, q^{1.46}$	0.89	23.8
Zn:	$c = 3.54\ 10^{-6}\, q^{2.79}$	0.99	51.7

Balances
The outcomes of the static balances of Cd, Cu, Pb and Zn in the farming systems are presented in Table III.3.6 (input (fertilizers and deposition) minus output (leaching and crop offtake) expressed in g ha^{-1} yr^{-1}).

Table III.3.6. Accumulation of Cd, Cu, Pb, and Zn in ecological (EAFS), integrated (IAFS), and conventional (mineral fertilizers only: MF, mineral and organic fertilizers: OF) arable farming systems at Nagele experimental farm expressed in g ha^{-1} yr^{-1}.

	Cd	Cu	Pb	Zn
EAFS	1.66	103.5	80.9	464.5
IAFS	0.75	53.3	37.6	702
CAFS-OF	0.96	53.5	46.3	804
CAFS-MF	4.93	-24.7	32	55

With respect to long-term simulation, the dynamic balance approach was applied to Cd and Cu, which represent extreme cases in the conventional arable farming system (CAFS-total). The conventional system even shows Cu depletion; a typical result of field-scale analysis as opposed to national studies.

Soil bulk density and plough layer thickness may not be constant in time due to changes in organic matter content and input or output of soil particles by processes like erosion. This is recognized by the so-called dynamic soil composition balance approach (DSCB), which takes into account changes in soil composition while calculating the dynamic balance (Chapter II.4.5.3). Because these changes in soil composition are not known, they are not regarded in this analysis.

The values for the input rates (A: Tables 2 & 3) removal rates by harvest (U: with area-weighted mean values of the individual B values given in Table III.3.7), and leaching rates (L: with solute concentration based on the adsorption models in Table III.3.5) were substituted in the dynamic balance equations (Chapter II.4.5).

Table III.3.7. Cd and Cu uptake rate constants (B: 10^{-4} yr^{-1}) for different arable farming systems at Nagele experimental farm (are-weighted mean values).

	Ecological	Integrated	Conventional		
			organic fert.	mineral fert.	Total
Cd	5.92	10.43	9.07	7.99	8.40
Cu	3.39	5.31	5.78	5.19	5.42

With these values, the development of soil content, leaching and uptake in the conventional system (total) was projected.. Soil Cd content will exceed the Dutch reference value (defined as [0.4 + 0.007{C + 3H}] with mass % clay (L) and mass % organic matter (H), i.e., 0.6 mg kg^{-1} for this soil) and develop towards a steady-state value of about 1.7 mg kg^{-1}. The leaching rate will increase from about 0 to 0.9 $g.ha^{-1}.yr^{-1}$ while the uptake rate in the conventional arable farming system will increase from 0.8 to 4.9 $g.ha^{-1}.yr^{-1}$. The high (average) offtake rate will result in quality standards of some crops being exceeded (Moolenaar et al., 1997a). The development of soil Cd content, leaching and uptake in the conventional system (total) is shown in Figure III.3.2

Figure III.3.2. Development of Cd input and soil content, leaching and uptake rates in the conventional arable farming system.

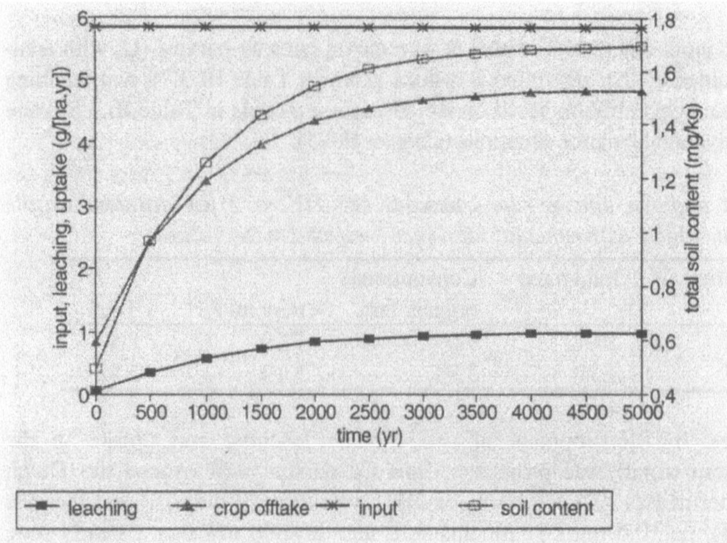

Copper soil content decreases from 60 to less than 14 mg.kg^{-1} at steady state, with associated reductions in leaching and crop offtake rates from 8 to 4 and 50 to 28 g.ha^{-1}.yr^{-1}, respectively. So, in the long run, Cd levels exceed soil quality standards whereas Cu is depleted. The development of copper soil content, leaching and uptake in the conventional system (total) is shown in Figure III.3.3.

Figure III.3.3. Development of Cu input and soil content, leaching and uptake rates in the conventional arable farming system.

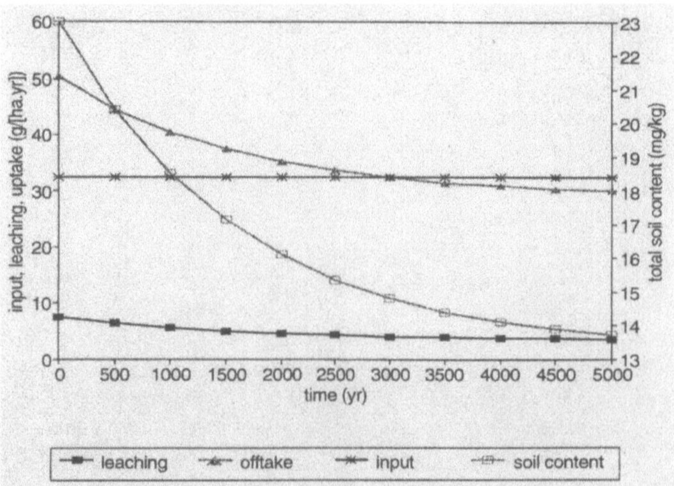

Discussion
These dynamic balance calculations illustrate that it is possible to compare and judge the long-term behavior of different heavy metals. Atmospheric deposition, crop rotation and selection of fertilizers directly influence both the annual balance and the long-term development of heavy-metal concentrations in soil, groundwater and crops. Only for CAFS-MF the contribution of atmospheric deposition to the total Cd input is lower than the contribution of fertilizer applications. The Cd balance of CAFS-MF shows a much larger accumulation than for CAFS-OF due to the Cd inputs with triple superphosphate applications. Copper and Zn inputs are highest when animal manures are applied and the contribution of atmospheric deposition only exceeds that of fertilizer in CAFS-MF.

Cultivation of crops with a high Cd offtake (carrots, sugar beets, ware potatoes and onions) also influences the Cd balance. EAFS has a higher Cd input/output ratio than IAFS and CAFS-OF because a larger percentage of the total area is grain crops with limited Cd offtake. Also inputs via manure are more significant in EAFS, partly due to high applications to raise the P status of the soil.

Clearly, the crop rotation and the selection of fertilizers directly influence the heavy-metal balance of arable farming systems. Optimization models may be used to formulate fertilizer plans that meet constraints on heavy-metal input via fertilizers, but also on other agricultural, legislative and economic constraints, based on farm-specific information. Velthof et al. (1996) using such an optimization model for arable farming indicated that it is not possible to formulate fertilizer plans in which input and output balance for Cd, Cu and Zn concurrently. Minimizing Cd input increases Cu and Zn inputs and minimizing Cu and Zn inputs increases Cd input due to substitution between animal manure and mineral fertilizer. This is also shown in the Nagele study. The integrated system compares favorably with the conventional (MF) system with respect to Cd, but the reverse holds for Cu, Pb and Zn, mainly due to different fertilizers used in these systems. In the conventional system, even Cu depletion occurs.

III.3.3 Case-study: Mixed farming systems

Mixed farming systems (Figure III.3.4) are very similar to dairy farming systems except that there may be output of roughage (grass and maize) and arable crops. In contrast to arable farming,, manure is mostly an internality (part of the internal flows) in mixed farming systems. Therefore, using a farm gate balance, closed loop accumulation cannot be detected in the case of arable farming whereas it may occur in the case of (dairy and) mixed farming.
By mixing land-bound farming systems (*i.e.*, arable and dairy farming), substance cycles could be closed by the exchange of forage crops with manure, and crop rotations could be widened which would result in less diseases and the potential for growing more profitable crops. Model studies indicate that this combination may improve both economic and ecological results (De Koeijer *et al.*, 1995). On experimental farm 'The Minderhoudhoeve' (near Swifterbant in the Flevopolder: 52° 33_ N, 5° 40_ E), two different mixed farming systems of this type are being developed: an integrated farm (135 ha) and an ecological farm (90 ha). Further details on the research plan of both mixed farms are provided in Lantinga & Van Laar (1997). We calculated heavy-metal

balances at the farm and field level based on the input, output, and internal flows as described in this research plan. Table III.3.8 shows the farm-gate balance of the integrated mixed farming system.

Figure III.3.4 Mixed farming system (dotted lines represent system boundaries).

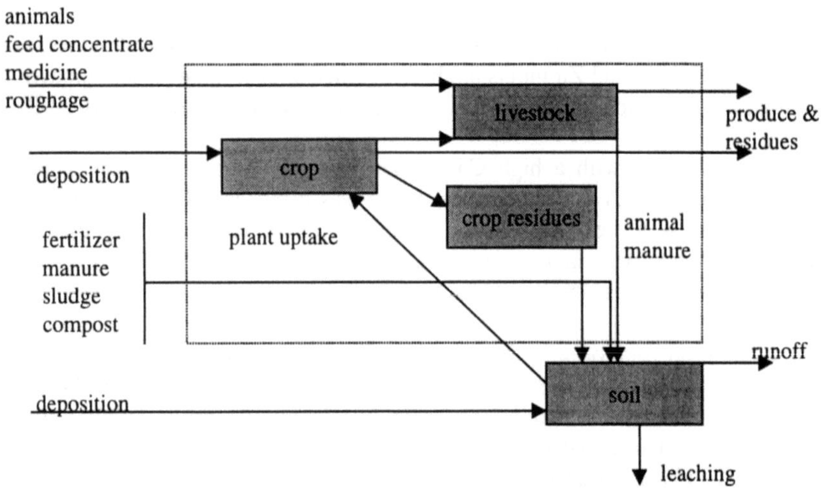

Table III.3.8 Farm-gate balance sheets of heavy metals (g ha^{-1} yr^{-1}) of the integrated mixed farming system at the Minderhoudhoeve.

	Cd	Cu	Pb	Zn
Input (*I*):				
Mineral fertilizers	1.37	1.2	1.5	16.2
Feed concentrates	0.05	16.5	0.5	52.8
Deposition	1.30	25.6	33.1	156.0
Total	**2.72**	**43.3**	**35.1**	**225.0**
Output (*O*):				
Crops	0.66	25.7	3.9	101.1
Milk	0.007	0.2	0.03	21.6
Meat/animal	0.0007	0.4	0.002	6.4
Leaching	1.6	43.1	14.8	63.3
Total	**2.3**	**69.4**	**18.7**	**192.4**
I* - *O	**0.42**	**-26.1**	**16.4**	**32.6**

The internal flows in the integrated mixed farming system consist of cow manure and roughage (grass, clover, fodder beets, silage maize). These flows are not part of the farm-gate balance, but they are important for the field-scale balances. On mixed farms, part of the land is grassland and part is used for growing arable and forage crops. The integrated farm comprises 41 ha grassland and 94 ha arable land. We derived separate field-scale balances for grassland and arable land. Grassland is amended with N-fertilizer and manure (52 ton ha^{-1}). The grass/clover mixture receives 18 ton manure per ha. Table III.3.9 shows the heavy-metal balances for the grassland.

Table III.3.9 Field-scale balance sheet (g ha^{-1} yr^{-1}) of grassland of the integrated mixed farming system at the Minderhoudhoeve.

	Cd	Cu	Pb	Zn
Input (*I*):				
Fertilizers/manure	2.4	219.1	89.6	876
Deposition	1.3	25.6	33.1	156
Total	**3.7**	**244.7**	**122.7**	**1032**
Output (*O*):				
Grass offtake	1.4	161.6	27.4	1378
Leaching	1.6	43.1	14.8	63.3
Total	**3.0**	**204.7**	**42.2**	**1441.3**
I - O	**0.7**	**40.0**	**80.5**	**-409**

Arable land is amended with about the same amount of N-fertilizer as grassland and with P-fertilizer and manure. Table III.3.10 shows the heavy-metal balances for the arable land.

Table III.3.10 Field-scale balance sheet (g ha^{-1} yr^{-1}) of arable land of the integrated mixed farming system at the Minderhoudhoeve.

	Cd	Cu	Pb	Zn
Input (*I*):				
Fertilizers/manure	2.1	95.4	38.7	387.5
Deposition	1.3	25.6	33.1	156.0
Total	**3.4**	**121.0**	**71.8**	**543.5**
Output (*O*):				
Feed crops	0.95	19.7	5.2	354.5
Arable crops	0.65	25.7	3.9	101.1
Leaching	1.6	43.1	14.8	63.3
Total	**3.2**	**88.5**	**23.9**	**518.9**
I - O	**0.2**	**32.5**	**47.9**	**24.6**

In the ecological mixed farming system, SSO-compost provides half the required P-input. The farm-gate balance of the ecological mixed farming system is shown in Table III.3.11. Almost all arable and forage crops are used for animal feed, hence crop offtake is restricted to 6 ha root crops. The field-scale balance of the ecological mixed farming system is shown in Table III.3.12. Internal flows consist of cow manure and practically all crops.

Table III.3.11 Farm-gate balance sheet (g ha^{-1} yr^{-1}) of the ecological mixed farming system at the Minderhoudhoeve.

	Cd	Cu	Pb	Zn
Input (I):				
Mineral fertilizers	0.03	2.7	3.0	8.9
SSO-compost	1.7	83	198	489
Concentrates	-	-	-	-
Deposition	1.3	25.6	33.1	156.0
Total	3.03	111.3	234.1	653.9
Output (O):				
Crop produce	0.36	17.6	1.4	253.8
Milk	-	0.18	0.02	19.3
Meat/animals	-	0.36	-	6.4
Leaching	1.6	43.1	14.8	63.3
Total	2.0	61.2	16.2	342.8
I - O	1.0	50.1	218	311

Discussion

Comparing Tables 8 and 11, shows that input with SSO-compost in the ecological system far exceeds the combined input with fertilizers and concentrates in the integrated system.

Tables 9 and 10 show for Cu, Pb and Zn in the integrated system that both the input in fertilizers and the output in crop products is (much) larger on grassland than on arable land. The balances show a larger Cd, Cu and Pb surplus for grassland than for arable land. On arable land, Zn accumulates and on grassland Zn depletion occurs.

For both the ecological farm and the integrated farm a discrepancy exists between the farm-gate balance and the total field-scale balance. Since the heavy metals do not degrade or volatilize, the difference must be visible in the livestock 'compartment'. However, the heavy-metal balance of the livestock compartment shows inconsistencies (not shown here). According to the calculations, Cd and Zn accumulate in the cows in relatively large amounts, while Cu and Pb seem to be 'produced' by the cows (Moolenaar & Lexmond, 1998).

Table III.3.12 Total (grass and arable land combined) field-scale balance sheet (g ha^{-1} yr^{-1}) of the ecological mixed farming system at the Minderhoudhoeve.

	Cd	Cu	Pb	Zn
Input (I):				
Mineral fertilizers	0.03	2.7	3.0	8.9
SSO-compost	1.7	83	198	489
Animal manure	0.6	100.0	40.0	400.0
Deposition	1.3	25.6	33.1	156.0
Total	3.63	211.3	274.1	1053.9
Output (O):				
Crop offtake (all)	1.5	100.7	13.6	999.1
Leaching	1.6	43.1	14.8	63.3
Total	3.1	143.8	28.4	1062.4
I - O	0.53	67.5	246	-8.5

This discrepancy between the internal flows from animal to crop and from crop to animal may have several causes:

- The literature based (*i.e.*, average) heavy-metal contents of concentrates and roughage are too low for Cu and Pb and too high for Cd and Zn;
- The literature based (*i.e.*, average) heavy-metal contents of animal (cow) manure are too high for Cu and Pb and too low for Cd and Zn;
- Significant Cu and Pb inputs may have been overlooked *e.g.*, from diffuse heavy-metal sources in extra feed additives, stable components and machinery, plumbing and water piping, soil ingestion, etc.

The heavy-metal contents in feedstuff and manure show large variations. Hence, it is risky to use averaged or literature values only. On-farm monitoring is needed to enable reliable on-site quantification. A combination of the possible causes for the discrepancies is most likely (Moolenaar & Lexmond, 1998). The uncertainties involved make it impossible to determine closed loop accumulation in a sound way for this case study.

III.3.4 Conclusions

General

Although the Dutch environmental policy supports the effect-oriented approach in the Dutch soil protection law, the path to soil quality standards that are closely related to effects will be long still. Therefore, a source-oriented policy is needed to limit heavy-metal fluxes that are known to be too large. The use of heavy-metal balance sheets as an extension of existing standards enables a principally different way of guarding soil quality. Instead of the (desired) status of the soil, the central focus point for analysis is the burdening of the soil with potential toxic elements per unit time and area (flux) and the subsequent net accumulation in the soil system.

The dynamic balance approach is a useful tool to compare the consequences of heavy-metal applications on agricultural land. Dynamic balance calculations can be carried out relatively simply if information is available about local application rates, soil and crop characteristics. The time-scale on which problems will occur is often directly related to heavy-metal application rates. Field-scale balances enable field specific and dynamic analysis of heavy-metal accumulation, leaching and uptake and consequently identification of 'hot spots' (e.g., specific fields, crops, applications, see III.3.3). Moreover, field scale analyses enable elucidation of the role of internal cycles and the quantification of soil-bound flows like dirt tare and erosion.

The difference between interpreting accumulation of essential and non-essential elements, fixation of metals in soil structures and the lack of suitable methods to quantify leaching from the plough layer, are important issues in heavy-metal balance research and at the same time part of an ongoing research and policy discussion.

Case study on arable farming

The development of soil content, leaching and uptake in the conventional system (total) was projected.. Soil Cd content will exceed the Dutch reference value and develop towards a steady-state value of about 1.7 mg kg^{-1}. The leaching rate will increase from about 0 to 0.9 $g.ha^{-1}.yr^{-1}$ while the uptake rate in the conventional arable farming system will increase from 0.8 to 4.9 $g.ha^{-1}.yr^{-1}$. The high (average) offtake rate will result in quality standards of some crops being exceeded.

Copper soil content decreases from 60 to less than 14 $mg.kg^{-1}$ at steady state, with associated reductions in leaching and crop offtake rates from 8 to 4 and 50 to 28 $g.ha^{-1}.yr^{-1}$, respectively.

These dynamic balance calculations illustrate that it is possible to compare and judge the long-term behavior of different heavy metals within different agricultural systems. Atmospheric deposition, crop rotation and selection of fertilizers directly influence both the annual balance and the long-term development of heavy-metal concentrations in soil, groundwater and crops.
Large Cd accumulation occurs if triple superphosphate is applied. Copper and Zn inputs are highest when animal manures are applied. Cultivation of crops with a high heavy-metal offtake influences the balances as well.

Minimizing Cd input increases Cu and Zn inputs and minimizing Cu and Zn inputs increases Cd input due to substitution between animal manure and mineral fertilizer. This is also shown in the Nagele study. The integrated system compares favorably with the conventional (MF) system with respect to Cd, but the reverse holds for Cu, Pb and Zn, mainly due to different fertilizers used in these systems. In the conventional system, even Cu depletion occurs

Case study on mixed farming
In the ecological system, input with compost exceeds the combined input with fertilizers and concentrates in the integrated system by far.

In the integrated system, both Cu, Pb and Zn input with fertilizers and output in crop products is (much) larger on grassland than on arable land. The balances show a larger Cd, Cu and Pb surplus for grassland than for arable land. On arable land, Zn accumulates and on grassland Zn depletion occurs.

A discrepancy between the internal flows from animal to crop and from crop to animal rendered the determination of closed loop accumulation impossible in this case study on mixed farming. If the system is defined on a national scale the outcomes may be quite different as was shown in III.1.

III.3.5 Recommendations for further research

Further dynamisation of balance models
Although data from long-term field experiments are needed to study the long-term environmental consequences of applying fertilizers and soil conditioners (cf. McBride, 1995), they may not give sufficient information because variation in data collection may hamper the reliability of data and it may be difficult to maintain relevance for current agricultural practices due to changing practices, technologies, cultivars and natural variation. Therefore, in addition to monitoring programs, projective models are needed to asses environmental consequences of different management practices. Dynamic heavy-metal balances are useful for projective purposes. The balance approach may thus be used in a strategy that advocates the prevention of future problems.

A fundamental constraint of the balance approach, however, is that dynamic balance studies generally consider the soil composition and the values of the input and output rate coefficients to be constant in time. Because the soil composition and properties may change in time, the values of the output rate coefficients may undergo significant changes in time as well. Also, the value of input rates is expected to change in time due to changing deposition, application rates, and composition of soil amendments. So, input and output rate parameters may vary to a large extent in both time and space and this variability should be taken into account in modeling of long-term heavy-metal behavior in the soil-plant system.

Further dynamisation of balance models would be made possible if requirements are met that enable the quantification of probability distribution functions of the input and output rate parameters. Improving system understanding and predictive ability requires integration of model development, field and laboratory experimentation, and performance monitoring of the system studied (Jakeman et al., 1993). Key variables that should be monitored could be gathered by systematic, sequential sampling over extended time periods using 'adequate' monitoring networks and 'representative' agro-ecosystems. Adequate means that reliable and standardized controls and analytical methods are used. Representative means, among others, that both comparable and different soil types are studied and that different agricultural practices are represented in the monitoring network.

IV.1 Introduction

Ester van der Voet & Lauran van Oers

In Section III a number of model calculations were performed for copper, zinc, lead and cadmium in the Netherlands. The answer to Research Question 3 of the 'Metals' programme, *what is the fate of the metals flowing into the Dutch economy*, has been found: one important sink appears to be the landfill of waste containing metals, but the major sink is the stock of products and materials circulating within Dutch society. This stock-building or accumulation implies that in future the generation of metal waste and emissions can be expected to rise again, unless specific action is taken. Calculations with FLUX, Dynabox and DSCB show that although at present human and ecosystem health standards are transgressed only incidentally, the present management regime does lead to a transgression of standards in the future, and is therefore not sustainable according to the definition adopted in the 'Metals' programme. This implies that changes must be made in the metals management regime in order to reach a sustainable state.

The steady-state calculations in Part III indicate only the ultimate result of the present regime, but do not specify how and when such an increase of environmental concentrations might be expected. The increase may be slow and steady, rising from the present level to the steady state over many years. Especially in the case of metal applications being phased out there may be a rather sudden and unexpected rise, leading at first to flows and concentrations far above the steady-state level, followed by a slow decrease to the steady state. Due consideration must be given to the fact that the stocks at present available in society are very large, i.e. several orders of magnitude larger than current annual emissions. In formulating policy, the issue of the implicit ultimate steady-state outcome must also be addressed and not only the prevention of undesirable situations in the shorter term." Zoals het nu staat is de tekst trouwens niet gekorrigeerd.

Part IV is therefore dedicated to the generation of solutions and the assessment of such solutions using the developed models. The analysis is limited to copper and zinc: for cadmium, the transgression of standards appears to be limited even in the steady state, and the lead problem is expected to be largely resolved by policy measures already taken, i.e. the phasing out of the addition of lead to petrol. Zinc and copper appear to be metals with applications still increasing. In this case problems are due both to relatively small emissions from very large societal stocks and to functional and non-functional trace applications in agriculture, both rather difficult to mitigate in an environmental policy.

Both the steady-state and the dynamic mode of the models are used to obtain a picture of the ultimate consequences of the policy packages included and of the time track towards that situation. An attempt has been made to use the results of all the developed models to arrive at policy recommendations. In the spirit of Section II.7, we do not aim at an integration of the models but at a subsequent application: one model answering the questions the others leave unattended. As in Part III, a combination of FLUX, Dynabox and indicators is used to build up a general picture of the situation in the Netherlands regarding copper and zinc. A more detailed result is obtained for the Dutch housing sector using FLUX. For Dutch agricultural soils, additional information and management guidelines are obtained with DCSB. The MPC models are used to derive recommendations regarding the possibilities for implementation and the optimum way to reach the desired changes in society. The aspects of costs and cost-effectiveness, however relevant for environmental policy, are neglected for the most part.

In Section IV.2 the origins of the metals problems in the Netherlands are discussed as a first step in the search for solutions. Subsequently, various management strategies are formulated, based on the assessment of origins. Such strategies as seem to have potential are then translated into scenarios. Three scenarios are specified to be entered into the models for assessment:

- a reference scenario based on not taking any action (of which the steady state has already been assessed in Part III)
- a 'moderate' scenario, including reasonable measures in agriculture and in the housing sector to gain insight into the solving power of relatively non-controversial options
- a 'stringent' scenario, which starts from the measures included in the 'moderate' scenario but adds as many as required to conform to the human and ecosystem health standards for environmental concentrations and daily intake. Depending on the outcome of the 'moderate' scenario, this scenario may also be rather moderate or it may be disruptive to society. In the first case, the message to policy-makers is optimistic: it is possible to solve the metals problem without too great an effort. In the latter case, we have a clear case of a confrontation between economy and environment, which cannot be readily solved and will cause policy-makers headaches in the near or more distant future.

In Section IV.3 the results of the model calculations are presented and discussed. This leads to a general discussion on the usefulness and value of models in IV.4, as well as some conclusions regarding a sustainable metals management regime.

IV.2 The generation of solutions for the environmental problems related to zinc and copper in the Netherlands

Ester van der Voet & Lauran van Oers

IV.2.1 Origins of zinc and copper problems in the Netherlands

The analysis of Section III.1 shows that for both copper and zinc the main problems are to be found in agricultural soil and in aquatic ecosystems. Under the present management regime, large-scale transgression of health standards may be expected roughly within a decade for aquatic ecosystems and within a century for agricultural soils; in view of the extremely slow immobilisation rate in the environment, this can be interpreted as rather soon. In addition, the quantity of metals entering the waste stage may pose a management problem: although recycling is a good option for many applications, processing large amounts of copper and zinc gives rise to a considerable volume of metal-containing waste-treatment residues which may eventually leach into the environment. In the steady state, both waste generation and leaching from landfill sites are significantly greater than in the 1990 situation.

Directions for sustainable management of zinc and copper may be obtained from an analysis of the origins of the problematical flows and accumulations. Origins can be defined at three levels (van der Voet, 1996):

- the direct causes: immediate immissions into the compartment
- the contribution of the economic sectors
- the ultimate origins at the border of the system.

For *aquatic ecosystems*, the origins of copper and zinc pollution are stated in Table IV.2.1. The values in this table exclude leaching from landfill sites and seepage; the FLUX steady-state model calculates extremely high values when forcing mass balance. This is difficult to interpret in a way that makes sense. However, it may be regarded as a warning that in the very long run landfill may be an important source of dispersion.

For both copper and zinc, the main source of 1990 is the transboundary inflow via the rivers Rhine and Meuse. In the steady state, with a much higher immission into surface waters, this source appears to lose some of its importance, although it is still large. The assumption has been that the transboundary inflow remains constant; therefore the relative decline is due to the fact that inland sources have increased in importance. For copper, emissions from sewage treatment plants have increased in the steady state. This is due to the expected increase in the corrosion of copper water pipes. Agriculture is a minor source in 1990, but in the steady state its contribution has increased by a factor 4.

Table IV.2.1 *Origins of the immission of copper and zinc in surface waters in the Netherlands*

	copper 1990	copper steady state	zinc 1990	zinc steady state
Total immission, surface water (t/y)*	569	1052 + ??? landfill	2776	3584 + ??? landfill
Direct causes (%)				
transboundary inflow, Rhine/Meuse	68%	37%	73%	56%
effluent from sewage treatment	12%	21%	05%	04%
anti-fouling treatment of ships	09%	05%	00%	00%
corrosion from building materials	00%	00%	10%	24%
industrial emissions	03%	03%	04%	03%
dissipative applications	02%	17%	03%	02%
atmospheric deposition	02%	01%	02%	02%
runoff, soils	03%	16%	03%	08%
runoff, landfill sites	00%	very high	00%	very high
Sectors responsible (%)				
foreign countries	70%	38%	75%	58%
industry	08%	07%	08%	06%
households/building	10%	35%	12%	26%
transport	09%	08%	02%	01%
agriculture	03%	13%	03%	08%
Ultimate origins (%)				
transboundary inflow	70%	38%	75%	58%
import, metals + funct'l metal appl.	28%	61%	20%	38%
import, raw mat's + non-func.appl.	02%	01%	05%	05%

Another relatively important source in 1990 is the anti-fouling treatment of ships. No increase or decrease is expected here. Emissions from industrial sources are expected to rise somewhat, but this is dwarfed by the increase of sewage emissions. For zinc, corrosion from building materials such as zinc gutters and zinc-coated railings is one of the major sources in 1990 and is expected to rise in the steady state. For agriculture, the pattern is the same as for copper: a minor contribution in 1990, but a large increase in the steady state. Other important 1990 emissions are sewage treatment effluent, industrial emissions from a number of processes involving zinc, mainly from chemical and ore processing plants, and some dissipative uses in households (e.g. paint). These emissions show hardly any increase in the steady state, which implies that their relative importance decreases. The only industrial source showing a marked increase is the food industry, due to higher copper and zinc contents in agricultural products. For surface waters, however, this is a minor source.

Transboundary inflow from foreign countries cannot be regulated by a Dutch policy. The main sources to address are therefore inland sources, viz.:
- copper water pipes
- anti-fouling treatment of ships with copper-containing materials
- zinc building materials or building materials coated with zinc

and especially in the long term:
- emissions from landfill sites
- agricultural flows of both copper and zinc.

The origins of copper and zinc pollution in *agricultural soils* can be found in Table IV.2.2.

Table IV.2.2 Origins of the immission of zinc and copper in agricultural soils in the Netherlands

	copper 1990	copper steady state	zinc 1990	zinc steady state
Total immission, agricultural soils (t/y)	1024	1959	2287	3415
Direct causes (%)				
animal manure	79%	82%	79%	86%
phosphate fertiliser	11%	06%	01%	01%
compost	06%	10%	04%	03%
pesticides	01%	00%	03%	02%
atmospheric deposition	04%	02%	13%	09%
Sectors responsible (%)				
agriculture	96%	98%	87%	91%
foreign countries	03%	02%	10%	07%
industry	01%	00%	03%	02%
Ultimate origins (%)				
transboundary inflow	03%	02%	10%	07%
import, metal + functional metal appl.	90?%	93?%	85?%	90?%
import, other raw mat's + non-func.appl.	07?%	05?%	05?%	03?%

For both copper and zinc the main source for agricultural soils is animal manure. The rise in the immission in the steady state can be attributed mainly to an increase in the copper content of manure. This is due to the fact that animal manure is part of a closed loop: animals eat fodder, which is grown with manure, which is produced by animals eating fodder... etc. To this closed loop, the main addition from outside is copper and zinc intentionally added to the fodder for productivity reasons. Such a closed loop implies that even with relatively small additions – and the additions are indeed small compared to other zinc and copper flows – environmental stocks may rise above health risk standards.

For zinc, atmospheric deposition due to inland and foreign sources is also important; in the steady state a relative decrease can be observed because inland sources are not expected to rise and transboundary inflow via the air is assumed to remain constant. For copper, phosphate fertiliser, containing zinc and copper as a pollutant, is the second most important source. This source, too, is relatively less important in the steady state. Both zinc and copper pesticides are added to agricultural soil but contribute in a minor way. Compost is part of a closed loop as well (food production – human consumption – cookery waste – food production); although the magnitude of this cycle is much smaller compared to the animal manure loop, we also see a rise in the steady state, especially for copper.

To solve the problems related to agricultural soil, it is therefore important to reduce the input to the closed cycles, viz.:

- copper and zinc additive to fodder

and to a lesser extent:

- copper and zinc pesticides
- phosphate fertiliser.

The copper and zinc entering the *waste treatment stage* can be found in Table IV.2.3.
In 1990, the main flow of waste generated is associated with building and demolition. These involve functional applications of copper and zinc such as zinc gutters and copper wiring. In the steady state, the amount of waste generated for both copper and zinc has increased significantly. Building and demolition waste still predominates: for copper, the relative contribution has dropped somewhat, although the absolute quantity has increased, but for zinc even the relative contribution has increased. The second most important source for copper is discarded consumer applications (pens, zippers etc.). For zinc, it is industrial waste, mainly from the zinc-ore processing industry but also from a large number of downstream zinc-processing industries. These sources are not expected to increase, and the relative contribution is therefore less in the steady state. Consumer applications in various types of products come third for zinc; for copper it is the non-functional flows as a contaminant in iron and steel and in concrete. The contribution of these applications to the generation of waste has risen in the steady state. Finally, the generation of waste from road construction activities can be mentioned, which is minor in 1990 but increases in the steady state owing to a larger metals content in waste materials re-used in roads.

Table IV.2.3 *Origins of the generation of copper and zinc waste in the Netherlands*

	copper 1990	copper steady state	zinc 1990	zinc steady state
Waste balance				
Total amount of waste generated in NL (t/y)	50423	74012	42449	65623
Total amount of waste imported (t/y)	48828	48828	04968	04968
Total amount of waste exported (t/y)	58959	69559	26803	42530
Total amount of waste treated in NL (t/y)	40292	53281	20614	28061
Direct causes of waste generated in NL (%)				
building & demolition waste	90%	80%	76%	82%
contaminant in concrete/iron & steel waste	01%	04%	01%	00%
discarded consumer applications	08%	09%	09%	07%
waste water pipes & rail wiring	00%	02%	00%	00%
waste road materials	00%	04%	01%	02%
industrial waste	00%	00%	13%	08%
Sectors responsible (%)				
building sector	92%	86%	77%	82%
households	08%	09%	09%	07%
transport	00%	05%	01%	02%
industry	00%	00%	12%	08%
Ultimate origins (%)				
transboundary inflow	00%	00%	00%	00%
import, metal + functional metal applications	99%	96%	99%	99%
import, other raw materials + non-func. appl.	01%	04%	01%	01%

To reduce the generation of waste, the main source to address is therefore:
- building and demolition

and to a lesser extent
- consumer applications
- industrial sources.

The origins analysis provides insight into the sources of the pollution problems and is therefore a valid starting point in the search for solutions. However, it does not specify how society should adapt in order to achieve the necessary changes. In Section IV.2.2 below, various ways to reduce the respective emissions are discussed in a general fashion. In IV.2.3 these options are translated into three management scenarios for copper and zinc.

IV.2.2 Directions for solutions

By returning to the processes within the economy, a better idea of the underlying causes of the increased emissions can be obtained. It is important to note that the increase of emissions takes place despite the already quite substantial recycling rates. Apart from cadmium, all the recycling rates of the waste-treatment phase were at least 80% in 1990. The non-recycled fraction relates to both accumulations and flows of trace applications and to non-functional flows, with only the latter in the steady-state situation. These trace applications and non-functional flows are the major cause of diffuse emissions to the sensitive routes. So even with a further increase of recycling, there will be leakages due to the 'squandering' of the metals occurring in the economy.

What options are available to deal with these risks? In general terms, three main approaches seem possible:

- the input into the economy can be lowered
- the output can be delayed
- the output can be controlled.

The first main approach, lowering the input, can be achieved in two different ways:

- by improving efficiency: boosting the efficiency of individual economic processes, and recycling metals. Increased efficiency automatically leads to a reduced inflow in the case of copper and zinc, which are metals with an elastic supply. However, this is not the case for metals with an inelastic supply such as cadmium, which is mined only as a by-product of zinc ore (van der Voet et al., 1994; Guinée et al., 1997), or mercury as a by-product of natural gas (Instituut voor Europees Milieubeleid & Environmental Resources Management, 1996).
- by decreasing the volume of use, either by substitution or by phasing out. Substitution of functional applications, such as zinc gutters by plastic ones (Kandelaars and van den Bergh, 1996), or copper-based pesticides by organic ones, is of course possible. An assessment must be made of the net environmental benefit of the substitution. Phasing out or at least a severe reduction is conceivable for the addition of copper and zinc to fodder, for example. A reduction of non-functional inflows can be achieved by mining ores with lower contaminant concentration, for instance, or by substitution of carrier materials like fossil fuels.

The second main approach, delaying of the output, can be achieved by using the metals in products with a long life span. A case in point is the use of fly-ash in concrete. The building case study (Fraanje & Verkuijlen, 1996) presents the framework for the assessment thereof. This option may be significant because it means technological improvements for the other options can be awaited. On the other hand, such a delay could be regarded as a shifting of problems to future generations and therefore undesirable.

The third main approach, control of the output, can be achieved in two different ways:

- physico-chemical immobilisation of the waste flow; high-temperature vitrification of metals implies in part delay and in part immobilisation of the outflow;
- choosing a direct route to final sinks in substrate, bypassing the sensitive environmental routes, such as a redirection of waste flows to mines or to the ocean floor.

Such a control of output does not take place at present: no adequate immobilisation techniques exist and waste storage in safe sinks is not practised. However, in view of the non-degradability of the metals, this is the only real 'disposal route' and therefore indispensable for sustainable metals management.

These general directions for metals management then need to be translated into specific directions in order to arrive at relevant policy recommendations. The indicators and analysis of Part III.1 show that the increase of emissions takes place despite an already quite high efficiency of industrial processes and an already high recycling rate. This implies that formulating a policy in that direction may not be adequate. The origins analysis of Section IV.2.1 is quite in line with this: it shows that the main problems arise from either involuntary losses from well-managed bulk applications in the built environment, or from small-scale and non-functional applications in the agricultural sector. In both cases, an immobilisation policy is not applicable either.
In order to combat emissions of the first type, i.e. corrosion, two routes are available:

- reducing the volume of use: a major reduction or a complete phasing-out of current copper and zinc applications and replacement by other materials
- improving 'efficiency': reducing corrosion during the use phase by technical means.

A volume policy, if applicable from a functional point of view, may lead to problem-shifting due to the emissions and waste generation connected with the production, use and waste treatment of the substitute materials. This should be considered before making such a recommendation. Reduction of corrosion might be obtained by means of coating. Again the question is how the reduction of copper and zinc emissions compares to the added emissions connected with the coating process. Coating may even reduce the possibilities for recycling, because it adds impurities to the metals.

For combating the pollution of agricultural soils, even fewer possibilities present themselves:

- reducing the volume of use: a major reduction or a complete phasing-out of present dissipative copper and zinc applications in agriculture. Substitution might be contemplated for pesticides and fodder additives. Whether this is possible or desirable is still an open question. It may be that such measures lead to reduced efficiency of agricultural production, which may lead to problem-shifting if the agricultural sector should react by trying to improve production again by other means.

A reduction of waste generation can be achieved by substitution or phasing out of bulk applications in buildings. Increased recycling of metals in demolition waste hardly seems feasible, since the recycling rate is already quite high. Again, the shifting of problems due to use of alternative materials should be a concern.

Apart from technical considerations regarding the feasibility of the considered changes, there are also economic and implementation considerations: the costs of taking such measures should be compared to the benefits of not transgressing metal-related health standards and attention should be given to the economic impacts of both. Implementation issues relate to the question of what policy instruments should be used for enforcing the changes and how maintenance and control is to be guaranteed. All such issues, however relevant, are treated only in a marginal manner in this study, the main effort being devoted to assessment in terms of the flows and stocks of the metals under review.

IV.2.3 Three scenarios for copper and zinc management

In this section, scenarios for the management of heavy metals are defined along the lines of the considerations in Section IV.2.2. This implies that measures to combat problems will be focused on the problematical applications in the built environment and in agriculture. Three scenarios are defined, each from its own perspective:

- A reference scenario, which can be used as a basis for comparison for the other two scenarios, based on maintaining the 1990 management regime.
- A 'moderate' scenario, starting from the possibilities for changes in management of the societal system and opting for such measures as are assumed to be relatively easy to implement.
- A 'stringent' scenario, taking environmental health standards as a starting point and comprising such measures as required to achieve those standards.

Reference scenario

In most scenario analyses, the reference scenario is some sort of a 'business as usual' scenario, comprising no specific measures besides already agreed-upon policies, but including socio-economic forecasts relating to population growth, welfare growth and autonomous changes in the structure of the economy based on present trends. Here we have opted for another possibility, viz. maintaining the 1990 management regime unchanged. This leads to improbabilities if the scenario is to be interpreted as a forecast: population, welfare and economic structure are constant throughout time, and policies already agreed upon are not implemented. However, it guarantees comparability with the steady-state calculations, as presented in Part III.1. A reference scenario such as this specifies the time path towards the steady state. It should not be regarded as a forecast of any future situation, but simply as the time-specific consequence of the 1990 management regime. Thus, it can serve as a 'clean' basis of comparison for the (changes resulting from) 'moderate' or 'stringent' policies, i.e. a basis not obscured by – very uncertain – assumptions regarding autonomous developments and potential future policies. On the one hand, this approach limits the applicability of such a scenario analysis. On the other, though, it clarifies the picture because it provides insight into the problem-solving power of the measures as such. The reference scenario is the basis of the model calculations presented in Section IV.3.1.

'Moderate' scenario

In the 'moderate' scenario, the main problematical applications in the housing and agricultural sectors are addressed. The measures proposed are moderate in the sense that they are considered as not too disruptive for society. It is not the solution of the metals problem but societal acceptability that is the bottom line. The purpose of this scenario is to assess whether the environmental problems in question can be solved with a minimum of societal problems. The proposed measures are the following:

measures, 'moderate' scenario	**copper**	**zinc**
technical emission reduction from open applications in present and new buildings	water decalcification: 2000-2010, 10% reduction of corrosion	coating of gutters, roofs, fences, etc.: 2000-2025, 75% reduction of corrosion
substitution of applications in the built environment – new buildings only	no substitution; more efficient building leads to 1% reduction of new water pipes per year, starting 2000	gutters, roofs, etc.: 70% substitution by stainless steel or plastics
terminating the use of metal-based pesticides	complete substitution by non-copper pesticides	complete substitution by non-zinc pesticides
reduction of additions to fodder	reduction by 30% of P and Cu additions, leaving production unaffected	reduction by 30% of P and Zn additions, leaving production unaffected
reduction of fertiliser use	reduction by 10%	reduction by 10%
reduction of livestock	reduction by 10%	reduction by 10%

In order to reduce corrosion from copper water pipes, decalcification of water is assumed to be applied in the 'moderate' scenario. This is expected to lead to a 10% reduction of corrosion in 2010, built up over a period of 10 years. Corrosive zinc applications can be coated, and this option is implemented in the 'moderate' scenario calculations for the full 100% of both new applications and stocks already in use over a period of 25 years starting in 2000. The assumption is that coating of roofs, gutters and fences leads to an emission reduction of 90%. This implies that the waste flow after discarding will be somewhat larger. The open applications constitute roughly 85% of all zinc applications in the built environment. Not included are zinc applications in central heating, radiators, ventilators and window frames inside the house. In all, this means a reduction of total corrosion by 75%, achieved in 2025. Coating is assumed not to be applicable for corrosive copper applications in the present stock: this concerns mainly water pipes, which cannot be coated once in place.

Substitution of copper is not possible in electric wiring and communication systems. Replacement by other materials, especially stainless steel, is in principle possible in water and gas pipes, water-heating equipment, and small-scale applications in locks etc., but is at present not feasible for practical reasons and for reasons of high cost. Substitution of zinc in new buildings and fences is assumed to be possible for 70% of applications. Only in the case of road fencing and brass applications in houses can zinc not be substituted, since no practical alternatives exist at present. Some minor indoor applications are also not replaced, since they do not add much to the waste flow and alternatives are not of the same quality. These substitutions are assumed to take place from 2000 onwards until 2010, when the use of copper and zinc in new buildings is supposed to be terminated. We then have left the non-replaceable applications and the stock of copper and zinc in older buildings, slowly entering the waste stage long after 2010. A trade-off will occur from zinc and copper to the mining, production, use and waste treatment of other materials. In the case of stainless steel, some of this will show up in the analysis due to the copper and zinc contamination present in the iron ore. Most of these trade-off impacts are outside the scope of this analysis and are signalled only in a qualitative manner.

Pesticides containing copper are used in buildings, in ship-building and maintenance, and in agriculture. Zinc pesticides are used only in agriculture. We have not investigated the possibilities, but the assumption in the 'moderate' scenario is that these pesticides can be replaced by other, organic and degradable substances, and therefore that their use can be terminated completely. The decline will start in 2000 and is complete in 2010. Apart from the fact that the above assumption may be rather optimistic – zinc pesticides are among those used on a large scale, and the use of copper pesticides has distinct advantages, especially for ship anti-fouling – there is also the question of the trade-off impacts of the organic pesticides replacing the Cu- and Zn-based ones. Again, this is beyond the scope of this study and is signalled only.

A reduction of phosphate addition to fodder – containing small amounts of copper and zinc as contaminants – is assumed to be possible through substitution with enzymes. Jongbloed et al. (1998) show that the present level of zinc and copper additions are no longer rational and that a significant reduction of the addition of zinc and copper is possible even without leading to any loss of production. In the 'moderate' scenario we assume that a 30% reduction is applied, starting in 2000 and completed in 2010, and that this does not affect agricultural production. This measure therefore does not have any side-effects.

The amount of copper added to soils by way of fertiliser is relatively low compared to other sources. In the 'moderate' scenario the assumption is therefore that the current downward trend in the use of phosphate fertiliser continues, leading to a 10% reduction in 2010.

The final assumption, regarding a 10% reduction of livestock, is a translation of the present agricultural policy in the Netherlands, which is not based on considerations regarding metals but on nutrification problems.

'Stringent' scenario

The starting point of the 'stringent' scenario is environmental quality: this scenario must comprise such measures as are required to remain within human and ecosystem health standards. The actual measures it comprises depend on the results of the 'moderate' scenario. There are three possibilities:

- the 'moderate' scenario solves the problems completely; in that case, the 'stringent' scenario is no longer required
- the 'moderate' scenario solves the problems to a large extent, but not completely; in that case, the 'stringent' scenario takes the 'moderate' scenario as a starting point and adds some measures in the same spirit
- the 'moderate' scenario leaves the problems largely unaffected; in that case, we must consider a more drastic approach.

In Section IV.3 we return to the 'stringent' scenario after the results of the calculations concerning the reference and 'moderate' scenarios have been presented.

References

- Fraanje, P.J. & E. Verkuijlen (1996). Balansen van non-ferro metalen in de Nederlandse woningbouw in 1990. IVAM onderzoeksreeks nr.76, Amsterdam
- Guinée, J.B., L. van Oers & E. van der Voet (1997). Cadmium in the Netherlands, a special case? CML report 136, Leiden.
- Instituut voor Europees Milieubeleid & Environmental Resources Management (1996). Mercury Stock Management in the Netherlands. Background Document for the workshop "Kwik, uitbannen of beheersen?" Ministry of the Environment, 21 November 1996, The Hague.
- Jongbloed, A.W., J.D. van der Klis, Z. Mroz, P.A. Kemme, H. Prins & B.W. Zaalmink (1998). Vermindering van koper, zink en cadmium in varkens- en pluimveevoeders. Een literatuuroverzicht. ID-DLO rapport 98.006.
- Kandelaars, P. and J. van den Bergh (1996). Materials-product chains: theory and an application to zinc and pvc gutters, *Environmental and resource economics*, Vol.8 pp. 97-118.
- Voet, E. van der (1996). Substances from cradle to grave. PhD thesis, Leiden University, defended 28 May 1996.
- Voet, E. van der, L. van Egmond, R. Kleijn & G. Huppes (1994). Cadmium in the European Community: a policy oriented analysis. Waste Management & Research (1994) 12, pp. 507 - 526.

IV.3 Results of the scenario calculations

Lauran van Oers, Ester van der Voet, Evert Verkuijlen, Patricia Kandelaars, Jeroen van den Bergh, Simon Moolenaar & Theo Lexmond

In this section the developed models presented in Part II are applied to calculate the impacts of the scenarios as described in Section IV.2. In IV.3.1 the FLUX - Dynabox combination in its dynamic mode is applied to the flows and stocks of copper and zinc in the Netherlands. Section IV.3.2 is dedicated mainly to the issue of stock modelling, which appears to be quite complicated and also very important for dynamic modelling. In Section IV.3.3 MPC modelling is applied to a specific materials-product chain: zinc gutters. This serves as an illustration of issues of costs and implementation, which are ignored by FLUX. Finally, in Section IV.3.4, the D[SC]B model is applied to calculate the copper and zinc concentrations in agricultural soils resulting from the assumed changes.

IV.3.1 The three scenarios for heavy metals in the Netherlands: calculations with FLUX and Dynabox

Lauran van Oers & Ester van der Voet

In this section the effectiveness of the scenarios specified in IV.2.3 for managing the environmental problems related to copper and zinc in the Netherlands are estimated using FLUX and Dynabox. To this end the steady state associated with the modified management regime is calculated, adopting a procedure comparable to that of III.1. In addition, the dynamic modes of FLUX and Dynabox are used to calculate developments in flows and stocks over time. Owing to the long residence time of the metals in both economic and environmental stocks, changes in management regime are likely to continue to have an impact long after the measures are assumed to have been implemented. We therefore choose the years for comparison rather far into the future from a policy point of view: 2050 and 2100.

During the application of FLUX in its dynamic mode we encountered several difficulties regarding the modelling of societal stocks and their outputs. The chosen approach for modelling these stocks may not be adequate in every case. The modelling outcomes are influenced significantly by the choice of modelling approach. The results presented in this section should therefore be regarded as indicative, with not too much value being attached to the actual numbers. In Section IV.3.2 these difficulties are elaborated further: the various possibilities for modelling societal stocks are discussed and the influence on outcomes is shown by using the stock of copper water pipes as an example. It should be noted, however, that the steady-state results are robust, especially concerning the flows, i.e. the steady states that result from each of these possibilities are identical. Only the road towards the steady state differs, but since this road may take centuries it is relevant.

Reference scenario

As explained in IV.2, the reference scenario is a simple continuation of the present management regime, with no regard for likely or possible autonomous developments. The steady state associated with this regime is therefore identical to the steady state presented in Section III.1. In this section we use the dynamic models to specify the road towards that steady state. FLUX generates time series of values for each flow and stock in the economy. The emissions series from FLUX are entered into Dynabox, which then calculates a time path for the resulting environmental flows, concentrations and intakes. The results of the calculations for the reference scenario will not be presented separately, but only in comparison with the 'moderate' and 'stringent' scenarios.

'Moderate' scenario

The terms of the 'moderate' scenario are described in Section IV.2. As mentioned above, the aim of this scenario is to establish whether the metals-related problems might be solved by a relatively straightforward strategy involving little disruption. Below, we first treat the steady-state results, thereby focusing on the three main problems: (1) ecosystem health, measured as concentrations in surface water and in agricultural soil, (2) human health, measured as daily intake, and (3) generation of copper and zinc waste. We then go on to consider the results of the calculations using the dynamic models.
In Figure IV.3.1 the indicators for human and ecosystem health are presented for 1990, the reference steady state and the 'moderate' steady state.

Figure IV.3.1 *Indicators for human and ecosystem health.*

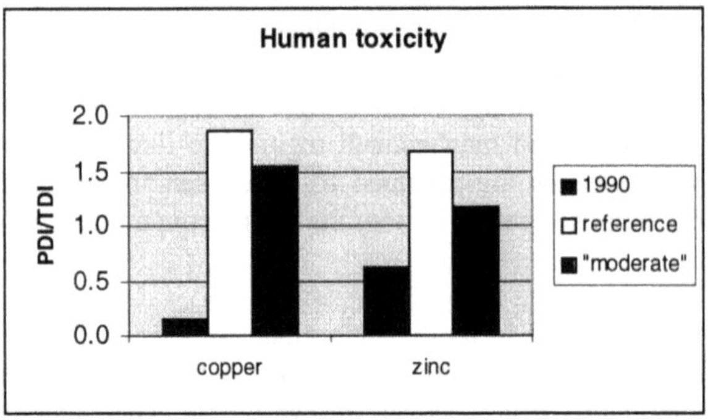

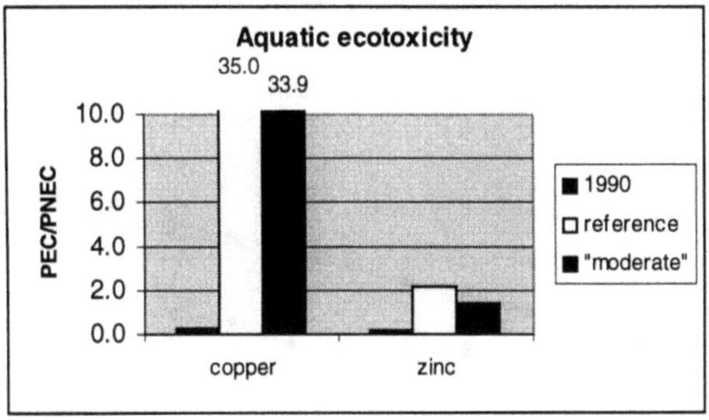

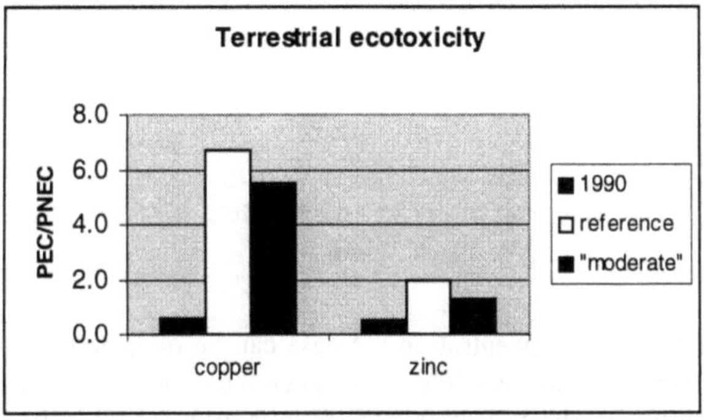

As Figure IV.3.1 shows, the 'moderate' scenario leads to lower concentrations and intakes in the steady state. However, the ADI/TDI and PNEC values are still transgressed, which implies that by our definition the 'moderate' management regime still is not sustainable. Since the transgression of standards is definitely less severe, especially for zinc, it may be the case that the length of time required to achieve the standards is far longer. This still does not make the scenario sustainable, but it may make the problem less urgent. This can be assessed with the dynamic models.

Figure IV.3.2 shows the development of zinc and copper concentrations in agricultural soils over the years, as calculated by Dynabox; these levels are directly relevant to terrestrial ecotoxicity and human intake.

Figure IV.3.2 Concentration of copper and zinc in soils compared to MPC.

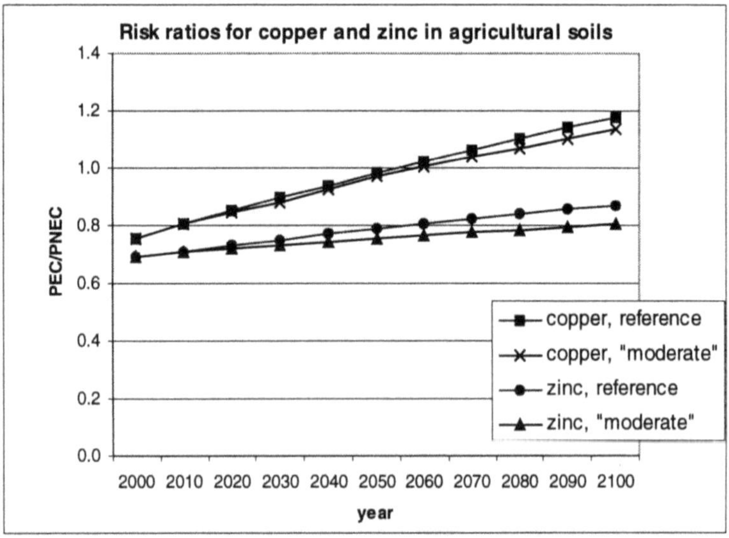

For both copper and zinc, a concentration increase can be observed over the years. In the 'moderate' scenario this increase is slightly less steep than under the reference management regime. This may influence the transition periods, although for agricultural soil this hardly applies. Table IV.3.2 below shows this.

Table IV.3.2 Transition period for risk ratios for copper and zinc in the Netherlands (years), reference scenario and 'moderate' scenario.

	copper, reference	copper, 'moderate'	zinc, reference	zinc, 'moderate'
MPC aquatic	years	years	years	decades
MPC terrestrial	decades	decades	a century	centuries
ADI	a century	centuries	a century	millennia

Although a considerable period of time may elapse before concentration limits are reached, this does not automatically imply that redemptive action can be postponed until then. The main lesson to be learnt from this modelling exercise is that changes in soil concentrations continue to occur for a very long time after changes in management regime. Figure IV.3.3 below shows this quite clearly, when compared to Figure IV.3.2 above.

Figure IV.3.3 Development of immission of copper and zinc into agricultural soils, 1990 - 2100.

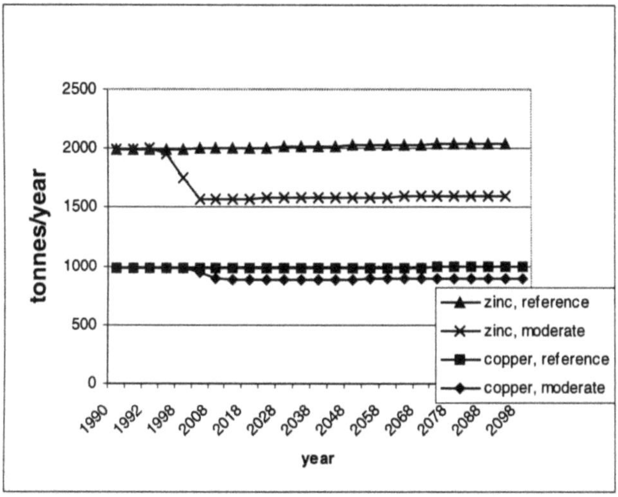

From this figure it can be concluded that the 'moderate' scenario already leads to a reduction of immissions into agricultural soils in the (relatively) short term, from 2000 onwards to 2010. This is the time period in which the measures are assumed to become effective. After that we see a very slow decline of the immission until 2100 for zinc, and a more or less steady level for copper. Comparing this to the concentration curves, we see a rise in concentration for both copper and zinc. In the reference scenario the input into agricultural soils is more or less constant until 2100. This does not mean that there is no change: since the output from soils due to harvesting and losses to the environment is lower than the input, the concentration still rises: comparing 2100 to the steady state it can be concluded that this increase will continue for centuries. After 2100 the immissions will rise again for both copper and zinc, ultimately leading to a significantly higher steady-state immission level. This can be explained by the closed-loop phenomenon: as concentrations build up, so does the amount of metals ending up in grass and fodder, and therefore also the concentration in manure, which ends up in the soil once again. Leaving the present metals management regime unchanged does not apparently mean that environmental quality remain similarly unchanged. Lowering the input still leads to rising concentrations, although more slowly. This gives rise to some considerations regarding environmental policy: although the lower input under the 'moderate' policy regime can be observed to have a certain beneficial impact, the concentration increase will not level off until more than a century after the measures have been implemented. This implies that policy actions aimed at keeping soil concentrations within standards cannot be taken too soon, and that more drastic measures may be required than would show up in a short-term analysis.

The most important conclusion from this modelling exercise regarding agricultural soils is that the 'moderate' scenario is not sufficient to solve the environmental problems. Inputs into agricultural soils must be reduced even more. For agricultural soils at least, a 'stringent' scenario must be defined.

Figure IV.3.4 shows trends in one of the main societal stocks stock of zinc: zinc in rolled and galvanised building materials in the reference and 'moderate' scenarios. Figure IV.3.5 shows the corrosion of zinc from that stock over time; this is a major source for aquatic ecosystems.

Figure IV.3.4 Stock of zinc in rolled and galvanised building materials, 1990 - 2100

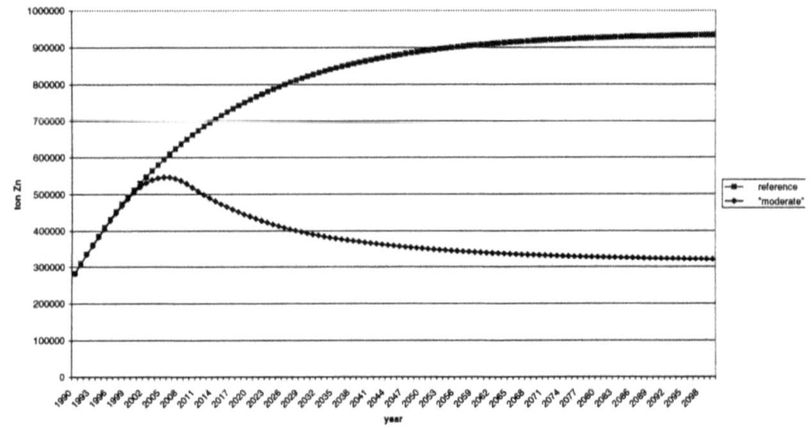

As might be expected, the two curves show great similarity. In the reference scenario, an increase of the stock by a factor 3 can be observed in 2100 compared to 1990. Although the stock is still increasing then, by 2100 the increase has more or less levelled off, leading to a steady state that is not much higher. The corrosion from the stock shows exactly the same curve. In the 'moderate' scenario the stock at first increases, similarly to the reference scenario. However, from 2010 onwards it decreases again, returning to the 1990 level again by 2100. This is due to the substitution of zinc by stainless steel and plastics. The corrosion is reduced even more because of the assumption regarding the coating of the remaining applications. After roughly 2035, the reduction of corrosion losses levels off. In 2100, the corrosion is way below half its 1990 level. This implies that at least one of the main sources of zinc for surface waters has been reduced significantly.

Figure IV.3.5 Corrosion of zinc from construction applications, 1990 - 2100

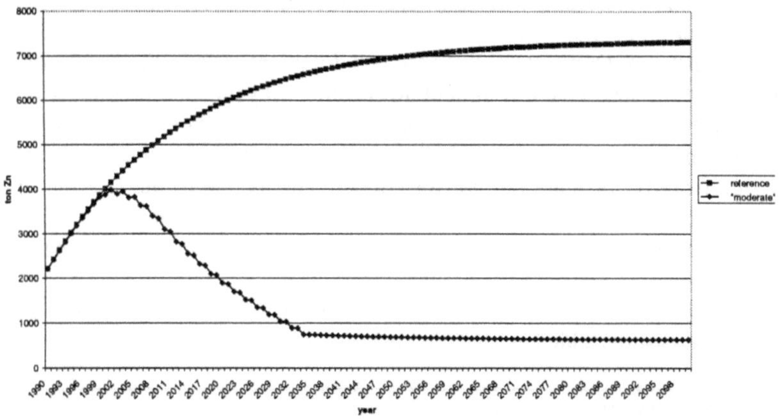

For the stocks of copper the difference between the reference and the 'moderate' scenario are much smaller, since most applications are considered non-substitutable for all practical purposes. This already shows in the health risk indicators: for copper, the reduction in the transgression of the standards in the steady state caused by the 'moderate' scenario is not very large and is due mostly to the changes in agriculture. Emissions to surface water are reduced somewhat by the assumption regarding decalcification of drinking water. This does not appear to make a very great difference for the steady-state situation.

For zinc, especially, the fact that a major domestic source has been reduced but standards are still transgressed leads to the suspicion that it might not be possible to clean up surface waters sufficiently with measures affecting only domestic sources. After all, the transboundary inflow via the Rhine and Meuse is also a major source of pollution, in 1990 as well as in the steady state. This suspicion was tested by running Dynabox with all emissions set to zero, and only the transboundary inflow left unchanged. This resulted in a steady-state risk ratio for aquatic ecosystems of 1.6 for copper and 1.5 for zinc. The conclusion therefore is that, although the transgression of the standards is definitely much less severe, domestic measures can never solve the problem of aquatic ecosystem pollution entirely, even in their most extreme variant. A 'stringent' scenario, therefore, although by necessity focusing mainly on domestic emissions, must contain assumptions about clean-up operations in foreign countries as well.

'Stringent' scenario

The 'moderate' scenario still leads to a transgression of environmental quality standards. More stringent measures are therefore required to solve the problems. In the case of aquatic ecosystems, foreign as well as domestic emissions must be tackled, since it is not possible to meet quality standards by national emission reduction alone. It is assumed therefore that the transboundary inflow will be reduced by 50% as a result of some sort of 'moderate' policy in other countries. In addition, domestic sources must also be reduced, especially for copper. For zinc, the major remaining source is leaching from sites and soils. This is also true of copper, but in addition corrosion from water pipes is still an important source. The input into agricultural soils must be reduced further to meet the standards for human intake and soil concentration. Domestic sources are still dominant even in the 'moderate' scenario. The main polluting sector is still agriculture itself. Stringent measures must therefore be defined for agricultural applications of copper and zinc. This should also have a beneficial impact on leaching from soils to surface water. Waste generation in the 'moderate' scenario is down significantly for zinc, but not for copper. This is due to the fact that the bulk of copper applications are considered irreplaceable in the 'moderate' scenario. This has consequences for both waste generation and leaching from sites to surface water.

The 'moderate' scenario is the starting point for the 'stringent' scenario, with a number of more stringent measures added as specified below:

Additional measures in 'stringent' scenario compared to 'moderate'	**copper**	**zinc**
Measures in foreign countries	'stringent' measures lead to 75% reduction of transboundary inflow by air and water	'stringent' measures lead to 75% reduction of transboundary inflow by air and water
Technical emission reduction from open applications in present and new buildings	no additional measures	no additional measures
Substitution of applications in the built environment; new buildings only	pipes and water-heating equipment: 100% of Cu substituted by stainless steel or plastic	additional substitution of Zn in road fencing: 80% substitution of Zn total
Terminating use of metal-based pesticides	terminate Cu-based anti-fouling	no additional measures
Reduction of additions to fodder to physiological minimum	reduction by 80% of Cu additions	reduction by 65% of Zn additions
Reduction of fertiliser use	reduction by 50%	reduction by 50%
Reduction of copper addition to soils	reduction by 75%	-
Reduction of livestock	reduction by 50%	reduction by 30%
Reduction of industrial emissions	reduced by 90%	-
Immobilisation of metals in landfill	100% reduction of emissions from landfill	100% reduction of emissions from landfill

No attention has been paid to the feasibility of the measures. Many of them may have an adverse societal impact. The agriculture-related measures certainly lead to a loss of production. Replacing copper with other materials in pipes and water heating is costly and leads to an increase of non-copper flows and stocks. Technical emission reduction in industries is only possible at high cost. A complete immobilisation of landfilled materials is not possible at present. It is required, nevertheless, in the 'stringent' scenario; given the nature of the steady state concept, it makes no difference whether the fraction lost is small or large, for in the steady state the losses automatically equal the yearly inflow. Therefore, two variants have been calculated: 'stringent 1' with no waste immobilisation, and 'stringent 2' with complete waste immobilisation.

When entered in FLUX and Dynabox, the measures lead to the steady-state results as presented in Figure IV.3.6 below.

Figure IV.3.6 Indicators for human and ecosystem health.

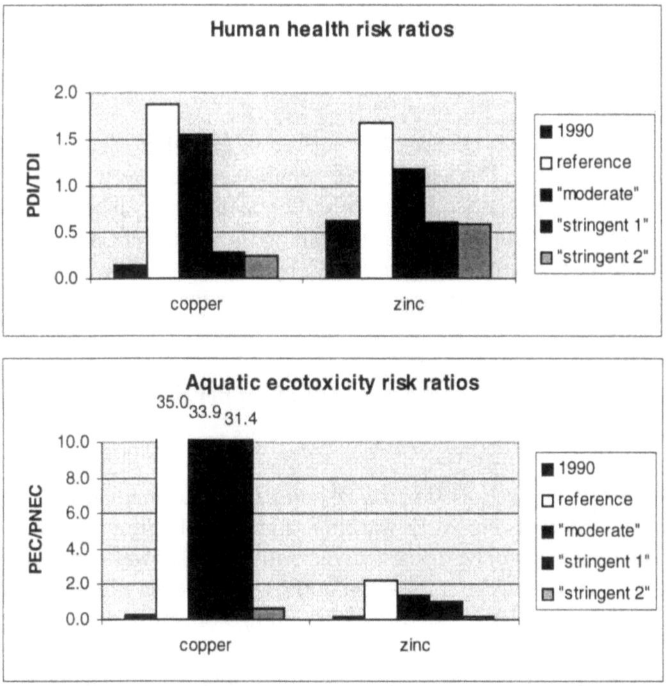

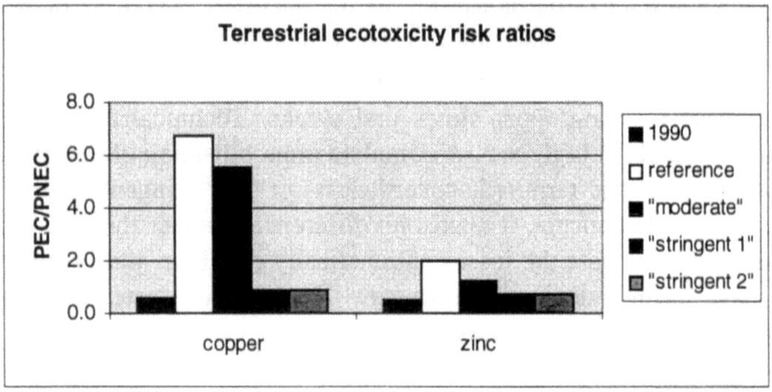

The overall conclusion from these calculations is that it appears to be very difficult to reduce the environmental flows and stocks of the metals to acceptable levels. The measures in the 'stringent' scenario are already quite unrealistic or even impossible at present. The side-effects and societal costs of such measures may be considerable. For copper even these measures do not appear to go far enough, although they do largely solve the zinc-related problems.

On a more detailed level, major changes are required in agriculture and in house-building to reduce emissions sufficiently. For aquatic ecosystems it appears very important to focus on waste management and especially on immobilisation techniques, since these are crucial for preventing leaching and runoff from landfill sites. Another major source is the transboundary inflow. This cannot be addressed by a national policy, but points rather at the importance of international harmonisation of metals policies.

IV.3.2 Dynamic modelling of stocks: the case of copper drinking water installations

Evert Verkuijlen, Ester van der Voet & Lauran van Oers

As already mentioned in Section II.6, stock-building or accumulation of metals in the societal system may serve as an 'early warning' signal for future emissions: one day, the applications of such metals will be discarded and will end up as waste and emissions. In the case of metals, this delay between inflow and outflow may be very long indeed; even if not used anymore, applications may still be kept in storage or left underground. Bergbäck & Lohm (1998) conclude that such "hibernating stocks" may be very large. In order to estimate future emissions, a crucial issue if environmental policy-makers are to anticipate future problems and take timely action, it therefore appears that such stocks cannot be ignored. It has recently been acknowledged that the main difference between static and dynamic SFA models lies in the inclusion of stocks in society: substances accumulated in stocks of materials and products in households or in the built environment (Bergbäck & Lohm, 1998; Baccini & Bader, 1996; Fraanje & Verkuijlen, 1996). Considering stocks is now a serious matter and has so far resulted in a number of exploratory ideas and a few specific substance stock models or databases (Lohm et al., 1997; Kleijn et al., 1998 in press). FLUX, too, contains a dynamic mode including stock-modelling, which is described in Section II.2 and has been applied to the case of metals in Section IV.3.1 above.

In order to control emissions in the long run, a stock management policy is required in addition to more 'source-oriented' policies. To formulate an adequate stock management regime requires due insight into the behaviour of societal stocks. The dynamics of such stocks depend on many variables. Among these are socio-economic aspects such as technological development, population size, welfare and market development. In interaction with these, substance, material and product characteristics play their role: these determine emissions during use, corrosion, life span, recycling potential and so on and so forth. An understanding of the mechanisms governing the generation of future emissions from present stocks of heavy metals, as a result of both socio-economic and geochemical aspects, is required before a sound stock model can be built.

There are basically two approaches to modelling the generation of emissions and waste flows from stocks. The first approach we typify, in accordance with Ishikawa et al. (in prep.), as the *leaching* model, the second as the *delay* model. The leaching model is the one used in FLUX and is quite straightforward: the generation of waste and emissions can be modelled as a fraction of the present stock size. An emission or discarding coefficient is defined and entered into the substance flow model. Such a coefficient can be based on empirical data, i.e. the waste flow or emission divided by the stock in a reference year. Alternatively, it can be based on information concerning the life span of the application: the reciprocal of this life span may then serve as a coefficient.

The delay model starts from the assumption that the output from societal stocks - the generation of waste and emissions - is determined by past input into and by residence time in the economy. The outflow in a certain year thus equals the inflow of a number of years earlier. This number of years is then the residence time. To build such a model, data are required on the present size as well as the historical build-up of societal stocks of substances, or – alternatively – on the stocks' inflows and outflows over past years. If available, such data might be used for building an empirical stock model. In practice such databases will be incomplete, and we must look for other ways to estimate stock behaviour. One possible approach to estimating such stock behaviour is to define stock characteristics (Ayres, 1978; Van der Voet et al., 1995). Available data on the build-up and size of stocks could then be used to validate the theoretical stock models and adjust their parameters if necessary. At present, this approach has been applied only as an example in a few specific cases (Olsthoorn, 1991; Kleijn et al., 1998). No attempt has yet been made to generalise it, much less to integrate it in a substance flow model.

Both approaches have their pros and cons. An obvious advantage of the leaching model is its simplicity. However, it may lead to inadequate results either through coincidents such as missing data, too rough stock estimates, the occurrence of a-typical years, or through the existence of stocks that are still in the build-up or already in the phase-out stage. The second approach, the delay model, does not have these disadvantages but requires much more insight into the build-up of stocks over the years. Another issue is the applicability of the two approaches. Theoretical considerations may dictate that some stocks, or some emissions from stocks, should be treated differently from others. For example, leaching from environmental stocks or corrosion from societal stocks may be modelled most adequately by using simple coefficients, as in the *leaching* model, since the actual metal molecules leaching out first are not necessarily the ones first entering the stock. Once in the stock, each molecule has an equal chance of leaching out. For the discarding of products, on the other hand, it may be more appropriate to use the delay model, since products obviously have a residence time after which they enter the waste stage.

In the case of applying the dynamic mode of FLUX to metals in the Netherlands, we are confined to the first approach. In view of the above, this may not be appropriate for all the stocks considered. To assess whether the results would differ dramatically, we modelled a single stock conforming to the second approach: copper in water pipes. The results are presented below, and are compared with 2 variants of the first approach: the empirical coefficient, and the lifespan-based coefficient.

Estimate of copper stocks in water pipes according to the leaching model

The technical lifetime of drinking water installations is estimated at 40 years (Fraanje & Verkuijlen, 1996). This results in a discarding coefficient of 0.025, which is entered into the leaching model with lifespan coefficient. On the other hand, from the estimated stock of 99 ktonnes and an outflow of 4.2 ktonnes a corresponding coefficient of 0.042 is calculated, which is entered into the leaching model with empirical coefficient. The difference in these coefficients is due mainly to the age distributions of dwellings. Figure IV.3.7 shows the results. From this figure we may conclude that the two possibilities within the leaching model - the lifespan-based coefficient and the empirical coefficient - already lead to quite different stock estimates.

Estimate of copper stocks in water pipes according to the delay model

Obviously, the delay model is more appropriate in the case of copper water pipes. Below, we try to obtain an impression of the differences in the outcome of the two models - although in theory the delay model is more credible, it may be that in practice the differences are not very large.

Figure IV.3.7 Estimates for stocks of copper in drinking water installations derived from FLUX calculations based on (1) the empirical lifetime coefficient and (2) the lifespan-based coefficient.

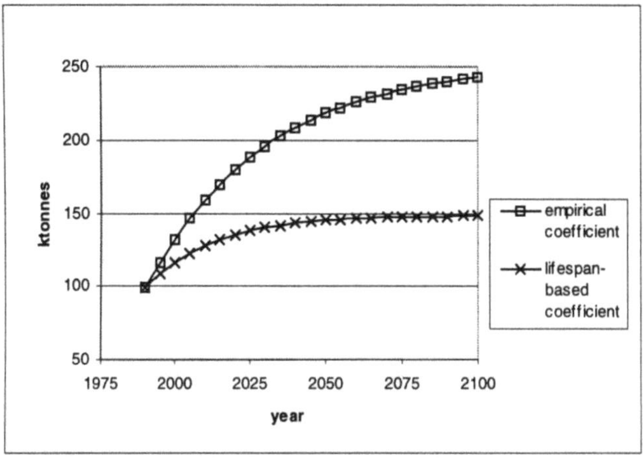

The calculated build-up of stocks according to the leaching model in the above is based on an estimated stock of 16.5 million dwellings in 2100. This estimate is also derived from the dynamic FLUX, and therefore on the leaching model. In view of the estimated population in 2100 of about 16 million, this figure is not very realistic. In practice, the outflow of copper from drinking water installations is not only determined by the lifetime of this application, but also depends on the number of houses that are renovated and demolished. Besides, installations are often adapted for alterations in kitchen equipment, sanitary, space and heating apparatus, etc. Connections to this kind of equipment have to be altered more than once during the life time of a drinking water installation as the specific life time of such equipment is 10-15 years. These alterations, too, result in additional in- and outflows of copper. Although difficult to estimate, Fraanje & Verkuijlen were able to express the volume of these alterations (and thereby the copper flows resulting from these alterations) in terms of the copper flows resulting from an equivalent number of dwellings in which the complete installation were to be replaced (Fraanje & Verkuijlen, 1997).

For a more adequate estimate of copper flows and stocks it is therefore more feasible to start from long-term data on the Dutch housing sector, that is, data on newly constructed and renovated houses in the past (in order to estimate future renovation and demolition), as well as data on future demand for dwellings. The latter can be derived from the long-term scenarios for the Dutch population and the housing sector.

Figure IV.3.8 Number of dwellings 1950 - 2100 based on long-term scenarios for population development and housing requirements.

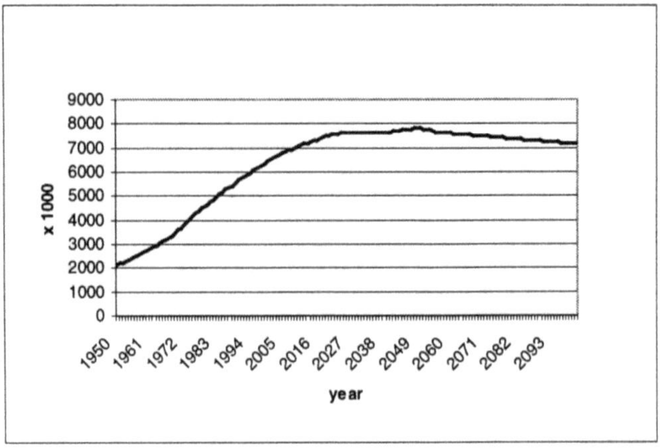

First, on the basis of these data, time series (1990 – 2100) were constructed of the number of dwellings to be constructed, renovated and demolished and the equivalent number of installations to be adapted (see Figure IV.3.9). The scale of these activities is subject to fluctuations owing to the large number of dwellings constructed between 1960 and 1980 and the growth of renovation activities between 1980 and 1995. As a consequence, the intensity of these activities will fluctuate in the future as well.

As a next step, the time series of the copper in- and outflows associated with these activities were derived. The inflow of copper in new water pipes was estimated on the basis of the average copper content in the reference year 1990. The outflow of copper in discarded water pipes was estimated according to:

$$F_i = N_{d,i} \,.\, M_{i\text{-life time } D} + N_{r,i} \,.\, M_{i\text{-life time } R} + N_{a,i} \,.\, M_{i\text{-life time } A}$$

wherein:

F_i = waste outflow in year i

$N_{d,i}$ = number of dwellings demolished in year i

$M_{i\text{-life time } D}$ = average copper mass per dwelling in year of construction/renovation of demolished dwellings in year i

$N_{r,i}$ = number of dwellings renovated in year i.

$M_{i\text{-life time } R}$ = average copper mass per dwelling in year of construction of renovated dwellings in year i

$N_{a,i}$ = equivalent of dwellings in which drinking water installations have been adapted in year i

$M_{i\text{-life time } A}$ = average copper mass per dwelling in year midway between construction and renovation

Figure IV.3.9 Activities in the housing sector regarding construction, demolition and renovation of dwellings and equivalent number of adapted drinking water installations (number of houses per year, 1950 – 2100).

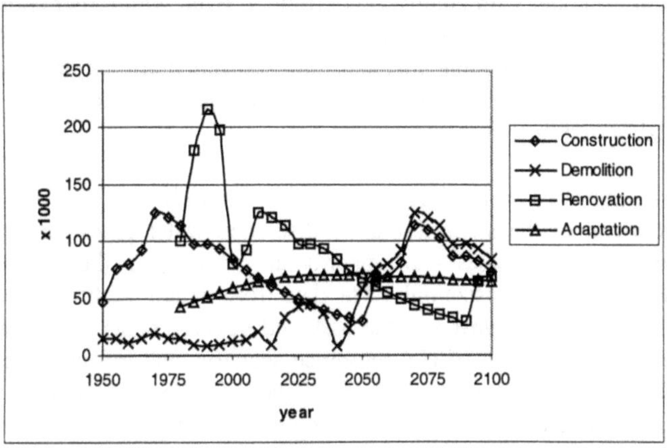

Figure IV.3.10 Estimates of copper stocks in drinking water installations of dwellings, 1990 - 2100, according to the leaching model (lifespan-based coefficient and empirical coefficient) and the delay model.

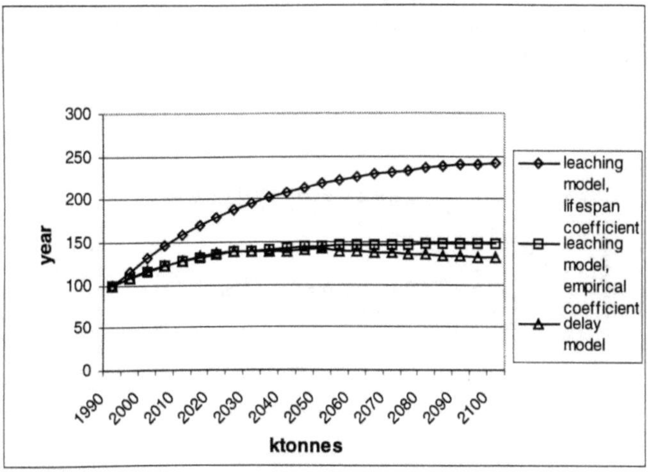

The size of the stock for a specific year can then be calculated from the previous year's stock and the present year's in- and outflow. Subsequently, in a stepwise process of calculating flows and stocks the whole time series is calculated. Here, dwellings are assumed to have a lifetime of 100 years and to be renovated every 40 years. The time series were constructed in a separate spreadsheet model. The results are presented in the following figure, Figure IV.3.10, and compared to the results of the *leaching* model.

The delay model leads to lower values for the stock's size than both variants of the leaching model. However, the actual values yielded by the delay model do not differ much from those of the leaching model with empirical coefficient. Even with rather similar stock sizes, the difference in the outflows (waste and emissions) may still be considerable. In the case of applications with a considerable lifespan, inflows and outflows are usually relatively small compared to stock size. As a result, the size of the stock is a fairly inert variable. Below, this is elaborated further.

Estimate of stock outflows with the leaching and delay models

The main outflow out of the stock is that of discarded water pipes, being transferred to either landfill or - in most cases - a copper-scrap refinery. In the leaching model, the outflow depends on the stock's size only. The data from which the time series is calculated is the stock's size in the base year and the assumed developments in inflow in the period 1990 - 2100. In the case of the FLUX reference scenario, the inflow is kept at the 1990 level of 6.3 ktonnes/year during the whole period. For the delay model, inflow developments are estimated based on the developments in the housing sector described above. The difference between the outflow curves therefore depends on (1) the assumptions regarding the inflow, and (2) the modelling assumptions regarding the generation of waste. Figure IV.3.11 shows the outflows of discarded water pipes in terms of copper, as calculated by the leaching and the delay model according to the above specifications.

Figure IV.3.11 Estimates of copper discarded from the stock of water pipes, 1990 - 2100, according to the leaching model (lifespan-based coefficient and empirical coefficient) and the delay model.

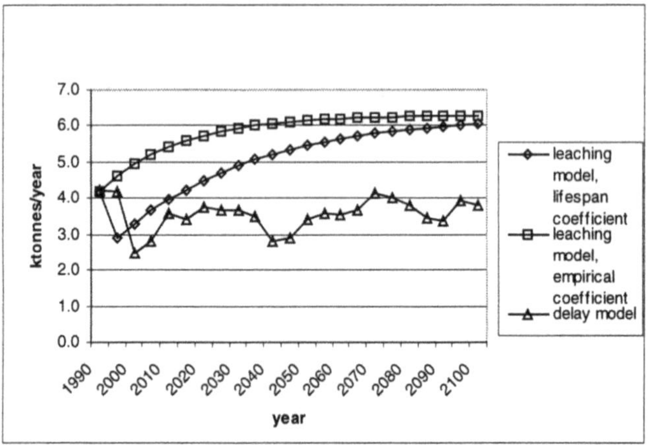

The difference between the outcomes of the two leaching models is considerably less than in the case of the stock's size, and decreases over time. The reason for this is that the outflow tends towards a steady state where it matches the inflow, which is identical in both cases (6.3 ktonnes/year, no changes over time). The delay model is quite a different story. Here there is no equilibrium to be reached. The line therefore develops differently and does not tend towards a certain value. In 2100, the predicted amount of copper entering the waste stage is only half the amount predicted by the leaching models. In this case it appears that it does make a large difference which model is chosen.

Another type of stock outflow is the corrosion of copper from water-pipes-in-use. As argued above, corrosion can be regarded as a leaching flow and therefore the leaching model is appropriate to use. A fraction is assumed of 0.35 g/kg copper stock per year. The differences in the curves of Figure IV.3.12 below therefore do not depend on the assumption regarding corrosion, but on the calculated size of the stock. The shape of the curves therefore exactly matches that in Figure IV.3.10 above.

Figure IV.3.12 Estimates of corrosion of copper from the stock of water pipes, 1990 - 2100, according to the leaching model (lifespan-based coefficient and empirical coefficient) and the delay model.

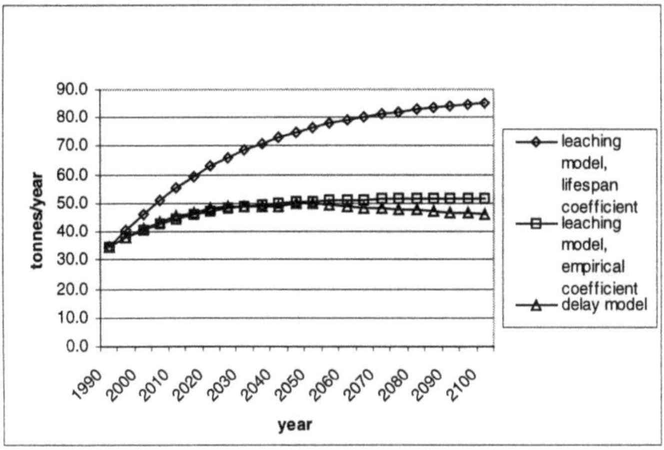

The moderate and stringent scenarios

In the above, the reference scenario was treated. Now we shall assess whether introducing changes in the inflow makes a difference. Under the moderate scenario, the assumption is that the inflow of new copper into the stock of pipes is reduced by 1% each year from 1990 onwards (see Section IV.2.3). This is entered into all three models. The stringent scenario has the assumption that copper in water pipes is phased out altogether. Over a 20-year period from 1990 onward the amount of copper entering the stock of pipes is reduced to zero. Figure IV.3.13 below shows the outflow of discarded copper pipes from the stock-in-use.

Figure IV.3.13 Estimates of copper discarded from the stock of water pipes, 1990 - 2100, according to the leaching model (lifespan-based coefficient and empirical coefficient) and the delay model, in the moderate and stringent scenarios.

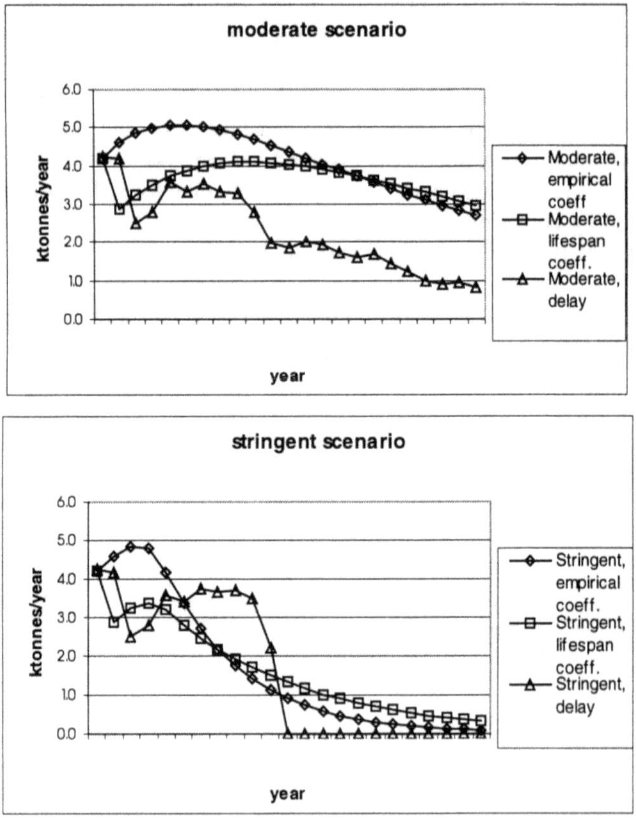

The differences between the delay and leaching models appears to be larger than ever. Under the assumptions of the moderate scenario, the amount of copper waste predicted by the leaching models is three times higher than the amount predicted by the delay model in 2100. The stringent scenario shows a peak value between 2025 and 2035 and a decline to zero by 2050 for the delay model, while the discarding flow develops much more gradually for both leaching models and continues beyond 2100.

Figure IV.3.14 shows developments regarding copper corrosion from water pipes according to the three models. As mentioned above, the shape of these curves exactly matches that of the stocks' size.

Figure IV.3.14 Estimate of copper corrosion from the stock of water pipes, 1990 - 2100, according to the leaching model (lifespan-based coefficient and empirical coefficient) and the delay model, in the moderate and stringent scenarios

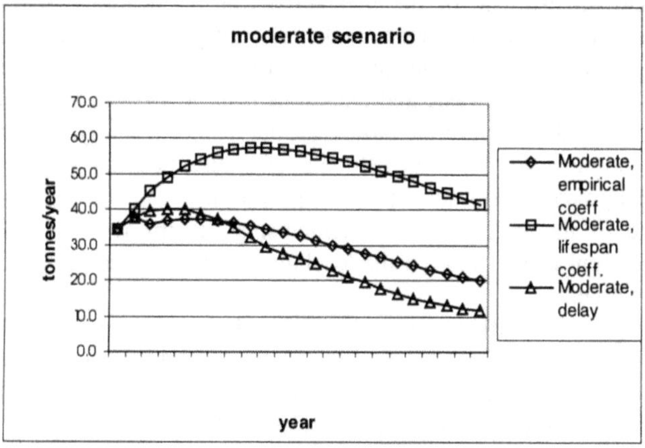

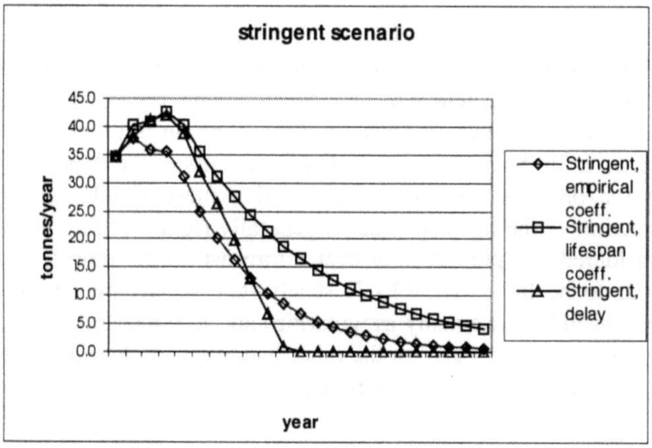

Under moderate conditions, the picture to emerge is rather similar to that for corrosion in the reference scenario: the delay model calculates the lowest values, but the difference with the leaching model using empirical coefficients is not very large. The difference between the two leaching models is much greater. The stringent scenario exhibits larger differences between the leaching and the delay approach. Here, corrosion is terminated by 2050 according to the delay model while it diminishes much more gradually under both leaching models, even continuing after until 2100.

Conclusions regarding the outcomes of FLUX in its dynamic mode
The calculations with FLUX as presented in the previous section were performed according to the leaching model with lifespan coefficient. This has the advantage of compatibility

with the FLUX steady-state calculations: at least we know that the road sketched by the dynamic FLUX is a road towards the same steady state. Moreover, it is a practical approach: in a way it makes sense, and data requirements are very limited (only the stock in the base year and an estimate of the lifespan of the application). However, from the example above it appears that in order to make relevant predictions regarding the stock of copper water pipes, use of a leaching model is insufficient. To say anything meaningful about the future generation of waste and emissions from this stock, knowledge is required not only of the size of the stock in the base year, but also of the way this stock is built up. Only then is it possible to make an estimate of future housing requirements and the corresponding equipment. Two issues appear to play a role:

- The estimate of the inflow curve. Harmonising the inflow curves leads to a less markedly different result between the leaching models on the one hand and the delay model on the other.
- The modelling of the outflow mechanism. This shows most clearly in Figure IV.3.13, especially for the stringent scenario. The delay model shows a termination of the outflow when the last inflow has lived out its lifespan, while the outflow from the leaching models in principle only reaches zero at infinity.

Whether these issues play a similar role in each stock cannot be determined out of hand. Clearly the problem is much less pronounced for applications with a short lifespan (< 5 years). The steepness and irregularity of past developments regarding the application may also be important. The dynamics of stocks in society are an interesting research topic on which very little work has been done as yet. For the results of the dynamic mode of FLUX as presented in the previous section, it implies that these should be approached with caution. The direction of the developments is generally right, but the outcomes should by no means be considered as a prediction or even a forecast. In the short term, especially, the differences appear to be large. The FLUX steady-state results may be considered more robust, on the one hand because many irregularities are no longer relevant - in most cases, not even the size of the stocks - and on the other because it is clear that the steady state is not a prediction but simply a basis for comparison.

IV.3.3 Contribution of Materials-Product Chain modelling to the scenarios

Patricia Kandelaars & Jeroen van den Bergh

M-P chain modelling aims, as described in Section II.3 and elaborated in Section III.2, at combining information regarding materials flows and market mechanisms. In view of the complexity of these models, it is not possible to design a comprehensive M-P chain model for all metals applications and the connecting flows. However, it is possible to apply M-P chain modelling to specific applications of the metals and thus arrive at additional recommendations and conclusions, in the manner described in Part II.1. In this section an M-P chain model is specified and applied for one application of zinc: gutters. It is a model of the type mentioned in Section III.2.4 under application C: a dynamic simulation model.

It is applied in addition to the FLUX-Dynabox combination for the 'moderate' scenario. This scenario, as specified in Section IV.2, contains a 70% reduction of zinc applications in the built environment. In this section, we investigate what is required to arrive at such a 70% reduction in the case of zinc gutters and what the consequences are, in both monetary and physical terms.

The demand for gutters is assumed to be given and met by two alternatives: zinc and PVC gutters. These two types of gutter form the basis of the model, which consists of eight sub-models: three at the product level, three at the material level, one for the extraction of ores, and finally one for the calculation of prices and costs. Figure IV.3.15 shows these eight sub-models and their links. Below, a general description is given of each sub-model.

Figure IV.3.15 The eight sub-models and their connections.

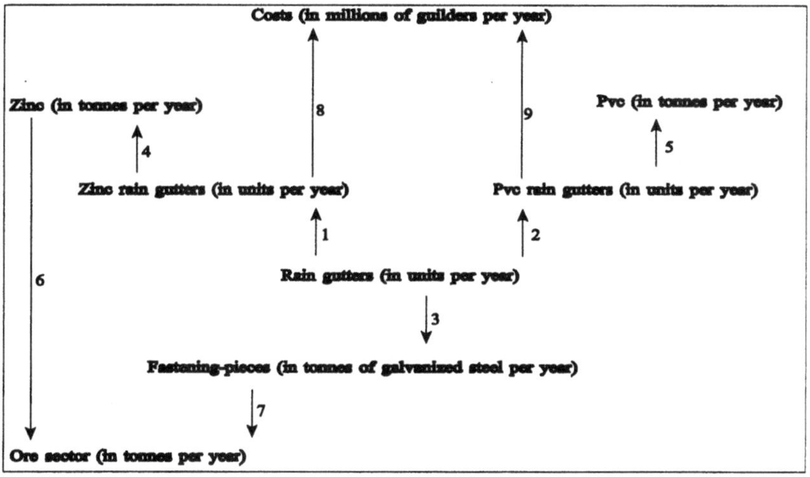

Demand for rain gutters

In this part of the model the demand, the accumulation in the economy and the waste of gutters are modelled. In the product sub-models, gutters are measured in functional units. In the model the total yearly demand for gutter services to be met is given exogenously. In the base year 1990 the number of gutters in the economy is 2,660,560. Every year some of these gutters are replaced and some are demolished, because the house to which they are affixed is destroyed. The replaced gutters and the gutters needed for newly built houses form the yearly demand for gutters. Gutters are not (yet) recycled at the product level.

Zinc and PVC rain gutters

These two sub-models describe the production of zinc and PVC gutters. The demand for gutters is allocated to zinc and PVC gutters (arrows 1 and 2 in Figure IV.3.15). This division of demand between the two types of gutter is one of the major decision variables of the model.

This division variable also determines the quantities of zinc and PVC needed to satisfy the demand for gutters. Obviously, total demand for gutters must be met by zinc and PVC gutters.

Fastening-pieces

A complete rain gutter service consists of fastening-pieces, made of galvanised steel. The fastening-pieces are measured in tonnes of galvanised steel for both types of gutters. The number and types of fastening-pieces made, demolished and renovated depend directly on allocation of the demand for gutters (arrow 3 in Figure IV.3.15). As opposed to PVC and zinc, galvanised steel is not recycled.

Zinc, PVC and galvanised steel

Here the quantities of zinc, PVC and galvanised steel are measured in tonnes. The zinc (PVC) sub-model is directly linked with the zinc gutter (PVC gutter) sub-model, indicated by arrow 4 (5) in Figure IV.3.15. The gutters (measured in functional units) are connected with the sub-models for zinc, PVC and galvanised steel by converting the number of products into the amount of materials. Conversion from functional units to tonnes is by means of a simple conversion factor. The materials needed for both types of gutter are calculated at the product level. Figure IV.3.16 illustrates the flows of zinc from the environment through the economy and back to the environment.

Figure IV.3.16 Disposal of materials in the environment as a result of flows in the economy.

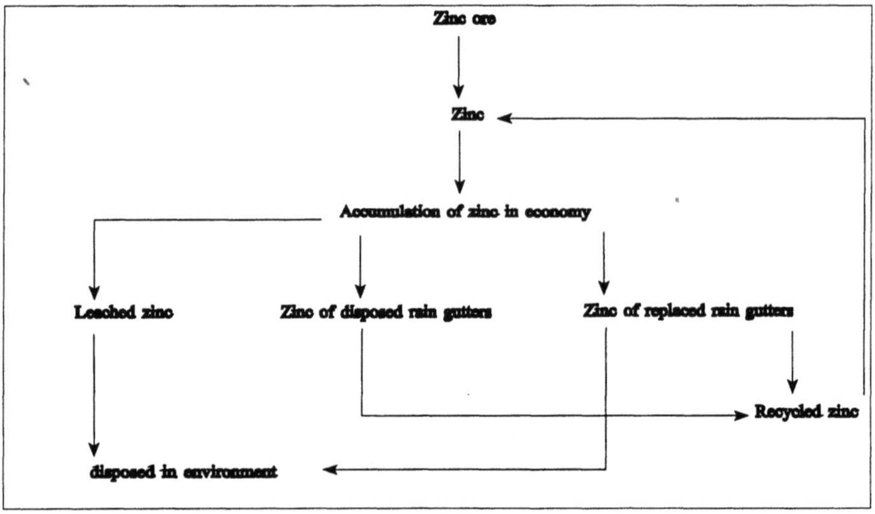

Ore sector

A sub-model for metal ores is added to describe and analyse the rate of extraction needed to meet the demand for gutters. Extraction may result in the depletion of zinc ore. Ores are needed for the production of zinc and of galvanised steel (arrows 6 and 7 in Figure IV.3.15), which means that the extraction rate depends on the demand for zinc and galvanised steel. Ores are measured in tonnes.

Prices and costs
In this economic sub-model the costs of meeting the demand for the M-P chain and the revenues of reusing materials are calculated. The costs are directly linked to the use of zinc and PVC gutters (arrows 8 and 9 in Figure IV.3.15). The costs involved are the prices of the two types of gutter and the revenues of used materials (zinc, PVC and galvanised steel). These revenues together with the prices of zinc and PVC gutters determine the allocation of demand between the two types of gutter. The net costs of a gutter equal the costs of buying a gutter minus the revenues of used materials. The kilogram price of new zinc equals that of recycled zinc, which makes it reasonable to assume that the difference between the price of recycled zinc and the revenue of used zinc equals the costs of recycling.

Simulation of the reference and moderate scenarios
The reference and moderate scenarios are analysed by way of dynamic simulation. These scenarios focus on the influences of preferential policy and economic change on material flows, through changes in M-P chains. In the reference scenario, the exogenously determined variables are held constant. The moderate scenario is 'implemented' by a product charge, which is adapted to reach 70% substitution of zinc by PVC. The control variables are: (1) the allocation variable that distributes demand for gutters between the two types of gutter; (2) the price of a zinc gutter, which influences the costs of meeting demand and the distribution of demand; (3) the price of recycled zinc, which influences the cost and the percentage and quantity of zinc that is recycled; and (4) the price of recycled PVC, which influences the cost and the percentage and quantity of PVC that is recycled.

For a comparison of the various scenarios the following clusters of performance indicators are distinguished: (1) the quantities extracted from and disposed of in the environment; (2) the allocation of demand for gutters; and, (3) the prices and net costs of satisfying gutter demand.

Reference scenario
In the base scenario all exogenous variables remain stable over time at the level of the base year 1990: demand for gutters, allocation over zinc and PVC gutters, recycling percentages for PVC, zinc and galvanised steel, dwelling demolition and gutter replacement. The data for the base scenario have been obtained from various sources mentioned in the model description in Kandelaars (1998, Chapter 8). Figures IV.3.17 to IV.3.19 show the results.

Figure IV.3.17 Demand for gutters and allocation over zinc and PVC, reference scenario.

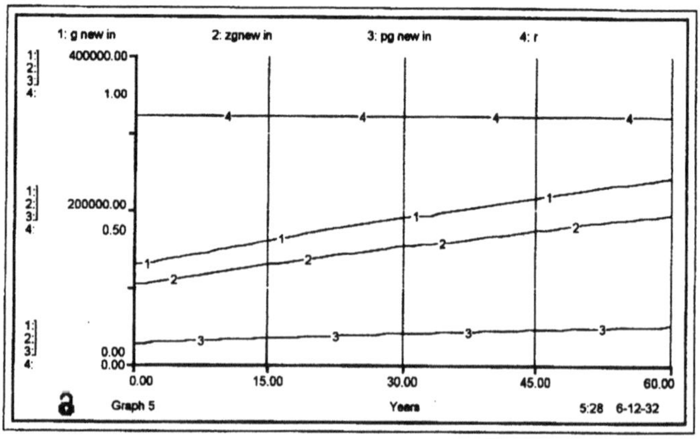

line 1: demand for gutters (pieces/year, 0-400,000)
line 2: demand for zinc gutters (pieces/year, 0-400,000)
line 3: demand for PVC gutters (pieces/year, 0-400,000)
line 4: allocation variable (fraction, 0-1).

Figure IV.3.18 Extraction and disposal of materials related to Dutch demand for gutters, reference scenario.

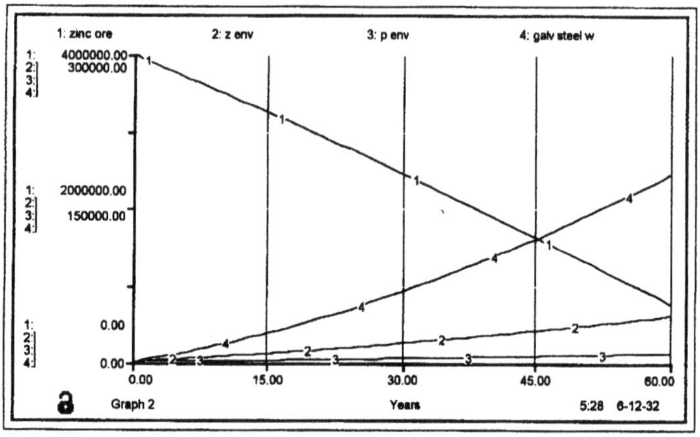

line 1: extraction of zinc ore (kg/year, 0-4,000,000)
line 2: disposal of zinc waste (kg/year, 0-300,000)
line 3: disposal of PVC waste (kg/year, 0-300,000)
line 4: disposal of galvanised steel waste (kg/year, 0-300,000).

Figure IV.3.19 *Prices of gutters and of recycled materials, net costs of meeting Dutch demand for gutters, reference scenario.*

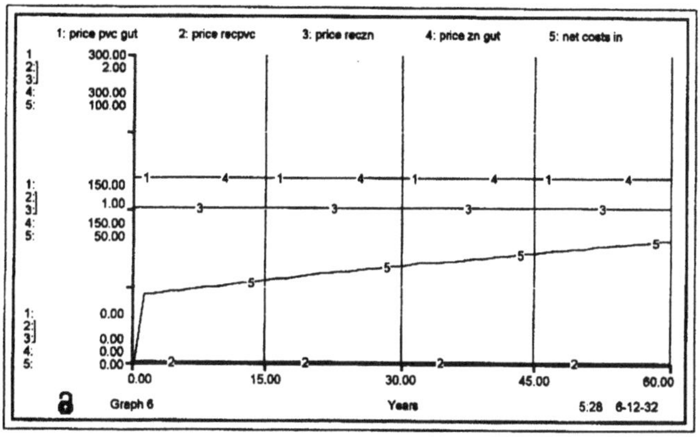

line 1: price of PVC gutters (Hfl/m, 0-300)
line 2: price of recycled PVC (Hfl/kg, 0-2)
line 3: price of recycled zinc (Hfl/kg, 0-2)
line 4: price of zinc gutters (Hfl/m, 0-300)
line 5: net costs of meeting demand (Hfl/m.y, 0-100).

Although the number of newly built houses per year is assumed to be constant, the reference scenario specifies a steadily increasing demand for new gutters over time (Figure IV.3.18), due to the fact that more houses are built than demolished. The allocation to zinc (80%) and PVC (20%) remains the same over the years. The extraction of zinc ore leads to a constantly decreasing stock of zinc according to Figure IV.3.18. The disposal of waste zinc, as well as that of waste PVC and galvanised steel, rises over time. Figure IV.3.19 shows that the prices of zinc and PVC gutters, as well as the prices of recycled zinc and PVC, remain constant as well. Because of growing demand, the net costs of meeting that demand rise over time.

Moderate scenario

This scenario differs from the reference scenario in that the allocation of demand depends on the prices of the two types of gutter. In the moderate scenario the price of zinc gutters is raised because of a charge that increases over time, as can be seen in Figure IV.3.22, starting in the base year 1990 and continuing until 25 years later in 2015. Figure IV.3.20 shows that the demand for new gutters still increases, although less than under reference conditions. However, the allocation to zinc and PVC shifts over time, resulting in a situation 25 years later wherein 70% of demand is met by PVC gutters and only 30% by zinc ones. This leads to very much reduced depletion of zinc resources according to Figure IV.3.21, although the disposal of waste materials has not changed very much. This would imply that the product charge is environmentally beneficial: environmental stocks are spared, while no alternative emissions take place. Figure IV.3.22 shows that the net costs in this situation are indeed lower than in the reference scenario, indicating that such a tax on zinc is a cost-effective measure to implement the desired reduction of 70%.

Figure IV.3.20 Demand for gutters and allocation over zinc and PVC, moderate scenario.

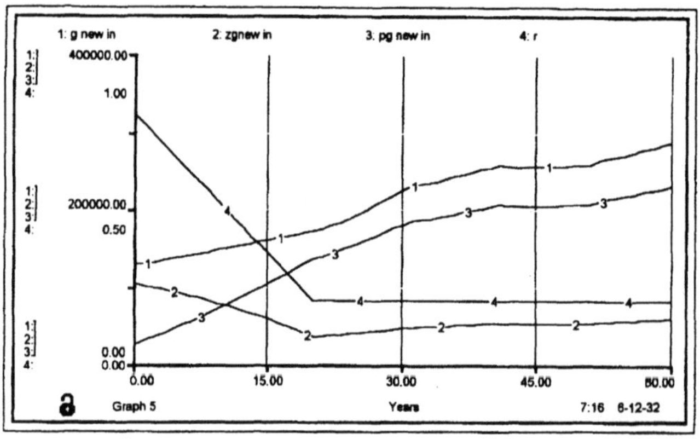

line 1: demand for gutters (pieces/year, 0-400,000)
line 2: demand for zinc gutters (pieces/year, 0-400,000)
line 3: demand for PVC gutters (pieces/year, 0-400,000)
line 4: allocation variable (fraction, 0-1)

Figure IV.3.21 Extraction and disposal of materials related to Dutch demand for gutters, moderate scenario.

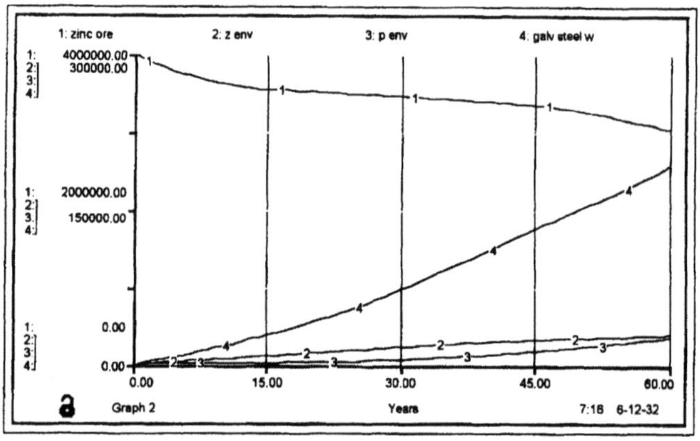

line 1: extraction of zinc ore (kg/year, 0-4,000,000)
line 2: disposal of zinc waste (kg/year, 0-300,000)
line 3: disposal of PVC waste (kg/year, 0-300,000)
line 4: disposal of galvanised steel waste (kg/year, 0-300,000).

Figure IV.3.22 Prices of gutters and of recycled materials, net costs of meeting Dutch demand for gutters, moderate scenario.

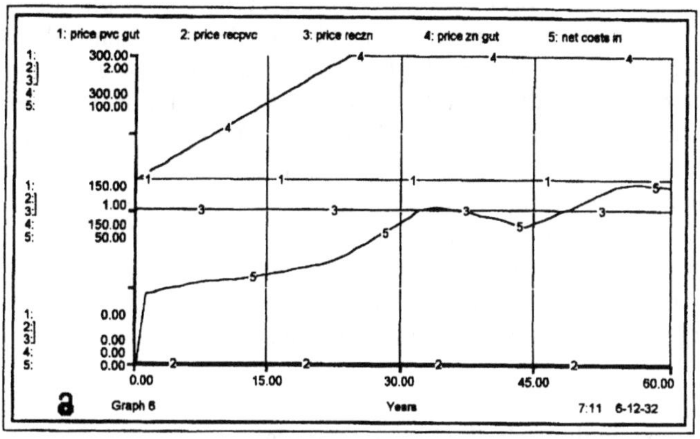

line 1: price of PVC gutters (Hfl/m, 0-300)
line 2: price of recycled PVC (Hfl/kg, 0-2)
line 3: price of recycled zinc (Hfl/kg, 0-2)
line 4: price of zinc gutters (Hfl/m, 0-300)
line 5: net costs of meeting demand (Hfl/m.y, 0-100).

From the application of this M-P chain model, various conclusions can be drawn:

1. The M-P chain model yields additional information that is valuable for policy purposes: it specifies changes in prices and costs due to use of a specific policy instrument (in this case, a tax) to implement the desired changes; it also specifies any consequent shifting of problems (in this case, a shift to PVC, which did not lead to an increase in emissions).
2. The application specified here is an example of how such a model can be used in combination with a substance flow model. There are many other instruments that can be included in such a model (see Kandelaars, 1998) and in principle such a model can be built for any application on the market.
3. The scope of the M-P chain model is limited to the markets it contains. Changes in zinc flows, unconnected to gutters, resulting from this product tax are not specified, nor is the influence of the tax on the recycling of zinc. Therefore, the best results may be obtained from a combination of the two models.

IV.3.4 Contribution of dynamic balances to scenario calculations

Simon Moolenaar & Theo Lexmond

The fate of the heavy metals Cd, Cu, Pb and Zn in agro-ecosystems is of concern because they are toxic to plants, animals and man. The control of heavy-metal fluxes is therefore one of the prerequisites for sustainable agricultural production. Agricultural cycles (soil - fodder - livestock - manure - soil and soil - food - human wastes - soil) are contaminated by heavy-metal inputs that may occur anywhere in the cycle, e.g. in the urban environment (corrosion, etc.), during transport processes and by supplementing animal feed. The use of waste products as soil conditioners and/or fertilisers thus may cause ever increasing heavy-metal flows (back) to agricultural soils on top of fresh inputs from mineral fertilisers, manure, atmospheric deposition, etc. In Chapter II.1.3 this mechanism is described as 'closed loop accumulation' causing the very high build-up of Cu and Zn in agricultural soils. The ultimate source behind the increasing Cu and Zn flows is animal fodder which is suppleted with Cu and Zn. As was shown in III.3, the D[SC]B calculations relate to (field and farm) scales. In IV.1 scenarios were presented that relate to the national scale. In this section the potential for using sustainability numbers (see II.4.7) based on dynamic balances for (large scale) scenario calculations are discussed.

Contribution of sustainability indicators to scenarios

The sustainability indicators presented in II.4.7 are based on dynamic balances and they enable screening and comparing of different agro-ecosystems without having to know all relevant processes in detail. Allowing for an assessment on a relative basis, these indicators reveal:

1. which heavy metal may cause the greatest violation of standards;
2. which environmental compartment is threatened most and for which compartment problems are expected first;
3. which experimental data assessment should have priority in view of decreasing uncertainty;
4. which approaches to avoiding violation of standards are feasible and most effective.

The proposed approach is valid for both non-essential elements (like Cd) and essential elements (like Cu and Zn). Both depletion and accumulation of heavy metals are related to sustainable agricultural practices and hence to the sustainability indices. However, for essential elements, appropriate standards for crop and groundwater quality may not always be defined. Instead, potential deficiency problems for crop production may be taken into account.

Closed loop accumulation

The scenarios based on FLUX and DYNABOX in III.1.3 and IV.3 do not show any depletion of the essential elements Cu and Zn at all. Rather, they seem to become the problem metals due to the mechanism of closed loop accumulation. The Cu and Zn flows are mainly related to purchased materials (feed concentrates and fodder) and the suppletion thereof. In this context, an important remark must be made. Crop uptake rates of Cu and Zn show wide variation and there is no fixed relationship between soil content and crop uptake. For Cu, a constant removal rate may sometimes be assumed since exclusion of Cu uptake may be operative (Mengel & Kirkby, 1982). Jarvis (1981) reported that over a wide range of total contents in a range of soils there is little relationship between Cu contents in soils and in plants. Van Luit & Henkens (1967) found that there was no further increase in Cu content in perennial rye grass, red clover and herbage after a Cu level of 5 mg/kg was reached in different humic sandy soils. These findings mean that it is questionable whether the supposed closed loop accumulation is actually taking place to the degree stated in III.3.1, because:

- plants will die as a result of phytotoxic Cu and Zn levels if these metals are taken up to any pronounced degree and therefore these plants will not become part of the fodder-animal-manure-soil-cycle;
- excluding mechanisms in certain crops may prevent Cu and Zn levels rising to the degree required for significant closed loop accumulation if they enter the fodder-animal-manure-soil-cycle.

If closed loop accumulation does not occur at the rate foreseen in III.3.1, the metals will accumulate in agricultural soils instead of being taken up by crops. Consequently, leaching rates may be significantly higher than those predicted in III.3.1.

Application of sustainability indicators in the scenarios

Regarding the contribution of dynamic balances to these scenarios, a study using the sustainability indicators on specific soils was carried out by Moolenaar et al. (1997a). This analysis proved to be very interesting for small-scale applications. However, results from basic research at small spatial (farm or field) scales must be related to larger spatial scales in order to address the questions raised from a more general policy and management point of view, as discussed in IV.2 and IV.3. Environmental management on the regional scale demands consideration of soil processes and the resultant fluxes occurring on that scale. Direct application of process-level information from a geographically smaller (e.g. field) scale to regional management is often problematic, and at times clearly in error. The methodology of scale translation therefore demands careful consideration. 'Down-scaling' studies are used to decompose process information (e.g. remotely sensed data) from the higher level to the lower (i.e. top-down) and 'up-scaling', or aggregation, studies use results from a smaller spatial scale to improve the understanding of processes at the regional scale (i.e. bottom-up) (Wagenet, 1996).

The input and output flows of heavy metals are a function of soil and management characteristics, which can be derived for the fields of a specific farm. For aggregation purposes, the question is how information about these soil characteristics is to be properly translated across different spatial and temporal scales. Field-scale analyses may be aggregated to a larger scale if it is assumed that the fields studied are representative of the farm, that the farm studied is representative of a specific farm-type, and that the farm-type represents a certain percentage of the agricultural sector in that region. This approach would result in application of a generic data set to a whole region. It is possible to reasonably estimate the input rate for different systems based on fertiliser, feedstuff and soil use. The values of crop uptake and leaching rates, however, are very site-specific. This complicates the aggregation of results at the field scale to farm and regional scales, and thus aggregation of field-scale balances could only be carried out if an extended database were available with distribution functions of the relevant input and output rates. Geographical Information Systems (GIS) enable an integrated assessment of environmental issues. If the information in GIS could be combined in such a way that these input and output rates were known, large-scale assessments might become possible. In that case, soils should be grouped according to their buffering capacity for heavy metals and the loads on these soils should be combined with these groups.

An option for linking the sustainability indicators of agro-ecosystems and the generic scenarios in III and IV may therefore be the incorporation of sustainability indices in a generic data set like a GIS. If the parameters in a GIS could be combined in such a way that the rate parameters (A, B and C) were known, this could be very promising for larger-scale (e.g. whole-region) assessments. As an intermediate approach, Guinée et al. (1999) used national statistical (averaged) data on different farming systems and fertiliser applications to quantify Cd flows and accumulation in the agricultural sector as a whole. An aggregated analysis at the national level was carried out by defining main crop-soil combinations that are representative of agriculture as a whole. For each combination, the input and output rate parameters were estimated and in this way a large-scale assessment could be carried out. Nonetheless, a well-defined large-scale dynamic analysis is impossible at this time for lack of the required data. For large-scale, static (one-year) assessments, existing national bookkeeping systems can be used. Different approaches to assessing heavy-metal balances are presented in Table IV.3.3.

Table IV.3.3 Different possibilities for assessing heavy-metal balances.

	Static	**Dynamic**
Top-down		
Large-scale	national bookkeeping system	not available
Small-scale	not applicable	not applicable
Bottom-up		
Large-scale	bookkeeping system (aggregated field/farm balances)	not available using D[SC]B & GIS?
Small-scale	field-scale (SB) balance & farm-gate balance; bookkeeping system	field-scale balance (D[SC]B)

SB: static balance; DB: dynamic balance; DSCB: dynamic soil composition balance; GIS: geographical information system

For the moment, the top-down and bottom-up results can be used in a complementary way as part of a research chain. With the former, investigations specific to sites and farming systems can be carried out, while general trends are discovered via the latter. This 'two-way street' involves up- and down-scaling. A challenging but important task for such a research chain would be to couple environmental effects with economic analyses such that the effects of heavy-metal management in agro-ecosystems could be stated in monetary terms as well.

Environmental management strategies

In terms of environmental management, it is important to discriminate between processes that can and that cannot be managed directly. Quality standards are based on policy choices. The crop production rate and the groundwater recharge rate are only partly influenced by human management (within economic constraints) and mainly determined by natural factors like soil type and climate. In the same way, the input rate is mainly determined by human influences, but can be only partially reduced by agriculture. The values of plant uptake and leaching rate mainly depend on natural processes, but may be influenced as well.

Short-term strategies

Different strategies for heavy-metal management will have different consequences for the resulting steady state. Short-term strategies aim at preventing problems in the short run, for example by increasing the soil's buffering capacity. Sorption characteristics vary with soil type and pH. For soils with a low sorption (or buffering) capacity, metal concentrations are expected to increase in groundwater and crops. This may result in groundwater and crop quality standards being exceeded and at the same time in little accumulation in soil.

'Good management practices' might attempt to lower the crop uptake rate, for example by raising the organic matter content (increasing retention) or by favouring competition by applying calcium and magnesium and competing heavy metals. Changing tillage practices may influence the stratification of pH, organic matter and metals. Moreover, cultivars may be changed and acid soils may be limed to increase the pH (Oliver et al., 1993, 1994; McLaughlin et al., 1994, 1995). Some practices that aim at lowering crop uptake (like stimulating competition) may at the same time lead to higher leaching rates or vice versa, thus resulting in a trade-off between leaching and uptake. Moreover, minimising output rates by management practices will result in the steady-state content being reached later at a higher level. Selecting cultivated crops with pronounced heavy-metal removal (within critical limits) can be very sensible for farming systems with low input. A higher uptake rate results in a lower steady state content (less soil accumulation) and higher crop offtake rates.

Long-term strategies

Long-term strategies focus on reducing inputs to soils to secure long-term soil quality. This results in the steady state being reached with lower total accumulation and lower output rates. Input reduction can be reached by reducing the amount of heavy metals in source material (quality) and by reducing the amount of fertiliser or manure added to the soil (quantity). This kind of input reduction could be sought by means of legislation, economic instruments, decreasing application and by educating farmers on how to use nutrient and heavy-metal balances.

The sustainability indices (II.4.7) meet some important criteria that have been identified by Gilbert and Feenstra (1994); they are representative for the chosen system, scientifically founded, quantifiable and represent the cause-effect chain for agricultural soils. Moreover, they indicate some points of action for heavy-metal policy and management, like setting up monitoring programmes to determine input, output and accumulation rates in agricultural soils. In this way, regional strategies for heavy-metal management can be developed that may be helpful in fine-tuning the 'moderate' (30% input reduction) and 'stringent' (80% input reduction) courses of action proposed in IV.2.

References:

- Ayres, R.U. (1978). Resources, Environment, and Economics, Applications of the Matcrials/Energy Balance Principle. John Wiley & Sons, New York / Chichester / Brisbane / Toronto.
- Baccini P. & H.-P. Bader (1996). Regionaler Stoffhaushalt, Spektrum Akademischer Verlag, Heidelberg Berlin Oxford.
- Bergbäck, B. and U. Lohm (1997). Metals in Society. In: D. Brune and V. Chapman (Editors.): The global environment - science, technology and management. Scandinavian Scientific Press, Oslo, pp. 276-289.
- Finkbeiner, M., M. Wiedemann & K. Saur. 1998. A comprehensive approach towards product and organisation related environmental management tools. Life cycle assessment (ISO 14040) and environmental management systems (ISO 14001). International Journal of LCA 3: 169-178.
- Fraanje, P.J., & E. Verkuijlen (1996). Balansen van non-ferro metalen in de Nederlandse woningbouw in 1990. IVAM-onderzoeksreeks no. 76, Amsterdam.
- Frosch, R.A. & N.E. Gallopoulos. 1989. Strategies for manufacturing. Scientific American 261: 144-152.
- Gilbert, A.J. & J.F. Feenstra. 1994. A sustainability indicator for the Dutch environmental policy theme 'Diffusion': Cadmium accumulation in soil. Ecological Economics 9: 253-265.
- Guinée, J.B., J.C.J.M. van den Bergh, J. Boelens, P.J. Fraanje, G. Huppes, P.P.A.A.H. Kandelaars, Th. M. Lexmond, S.W. Moolenaar, A.A. Olsthoorn, H.A. Udo de Haes, E. Verkuijlen & E. van der Voet. 1999. Evaluation of risks of metal flows and accumulations in economy and environment Ecological Economics (forthcoming).
- Ishikawa, M., R. Huele, R. Kleijn & E. van der Voet (in prep.). Predicting future emissions based on characteristics of stocks.
- Jarvis, S.C. 1981. Copper concentrations in plants and their relationship to soil properties. In: J.F. Loneragan, A.D. Robson, R.D. Graham (Eds.); Copper in soils and plants. Academic Press Australia, pp. 265-285.

- Kleijn, R., R. Huele & E. van der Voet (1998). Dynamic Substance Flow Analysis: the delaying mechanism of stocks, with the case of PVC in Sweden. Ecological Economics, in press.
- Lambert, A.J.D. 1994. Chain description of production systems. Milieu 1: 12-22.
- Lohm, U., B. Bergbäck, J. Hedbrant, A. Jonsson, J. Svidén, L. Sörme & C. Östlund (1997). Databasen Stockhome. Flöden och ackumulation av metaller i Stockholms teknosfär. Tema V Rapport 25. Linköpings Universitet. (In Swedish).
- McLaughlin, M.J., C.M.J. Williams, A. McKay, R. Kirkham, J. Gunton, K.J. Jackson, R. Thompson, B. Dowling, D. Partington, M.K. Smart & K.G. Tiller. 1994. Effect of cultivar on uptake of cadmium by potato tubers. Australian Journal of Agricultural Research 45: 1483-1495.
- McLaughlin, M.J., N.A. Maier, K. Freeman, K.G. Tiller, C.M.J. Williams & M.K. Smart. 1995. Effect of potassic and phosphatic fertiliser type, fertiliser Cd concentration and zinc rate on cadmium uptake by potatoes. Fertiliser Research 40: 63-70.
- Mengel, K. & E.A. Kirkby. 1982. Principles of plant nutrition. International Potash Institute, Switzerland, 655p.
- Moolenaar, S.W. 1999. Heavy-metal balances. II: Management of cadmium, copper, lead and zinc in European agro-ecosystems. Journal of Industrial Ecology 3 (1).
- Moolenaar, S.W., S.E.A.T.M. van der Zee & Th.M. Lexmond. 1997a. Indicators of the sustainability of heavy metal management in agro-ecosystems. The Science of the Total Environment 201: 155-169.
- Oliver, D.P., J.E. Schultz, K.G. Tiller & R.H. Merry. 1993. The effect of crop rotations and tillage practices on cadmium concentration in wheat grain. Australian Journal of Agricultural Research 44: 1221-34.
- Oliver, D.P., R. Hannam, K.G. Tiller, N.S. Wilhelm, R.H. Merry & G.D. Cozens. 1994. The effects of zinc fertilisation on cadmium concentration in wheat grain. Journal of Environmental Quality 23: 705-711.
- Olsthoorn, A.A. (1991). Sources of persistent micropollutants: analysis with dynamic materials balances. In: J.B. Opschoor & D. Pearce (eds.): Persistent pollutants: economics and policy. Kluwer, Dordrecht.
- Udo de Haes, H.A. & G.R. de Snoo. 1996. Environmental certification. Companies and products: two vehicles for a life cycle approach? International Journal of LCA 1: 168-170.
- Udo de Haes, H.A. & G.R. de Snoo. 1997. The Agro-Production Chain: Environmental management in the agricultural production consumption chain. International Journal of LCA 2: 33-38.
- Van Luit, D. & C.H. Henkens. 1967. Effect of the copper status of the soil on the copper content of grass and clover (in Dutch with English and German summary). Verslagen van landbouwkundige onderzoekingen. Centrum voor landbouwpublikaties en land-bouwdocumentatie Wageningen, 33 pp.
- Voet, E. van der, R. Kleijn & G. Huppes (1995). Economic characteristics of chemicals as a basis for pollutants policy. Ecological Economics 13 pp 11-26.

- Wagenet, R.J. 1996. Description of soil processes and mass fluxes at the regional scale. In: A.A.M. Del Re, E. Capri, S.P. Evans & M. Trevisan (Eds.); The environmental fate of xenobiotics. Proceedings of the X symposium pesticide chemistry, September 30-October 2, Piacenza, pp. 1-18.

IV.4 Discussion and conclusions

Ester van der Voet & Lauran van Oers

In this section, the main results of the model calculations for the three scenarios are discussed. In Section IV.4.1 the value of the various models is discussed in terms of their contribution to evaluation of management of copper and zinc and to the design of a 'sustainable' management regime for these two metals. Section IV.4.2 considers such a sustainable regime, assessing the suitability of various options for management changes.

IV.4.1 Discussion and conclusions regarding use and value of the models

Two models have been used to evaluate options for overall Dutch metals management: FLUX and Dynabox. Both are essentially SFA models.

FLUX (see Section II.2 for a description) has been applied to the flows and stocks in the Dutch economy, Dynabox to the environmental flows and stocks. In evaluating management options, FLUX has been used for the following purposes:

- to inventory flows and stocks in, out and through the economy for the base year 1990
- to calculate steady-state societal flows and stocks, the 'ultimate consequence of the present regime'
- to compare alternative regimes with the present regime on a steady-state basis.

The outcomes of FLUX have been translated into sustainability indicators regarding the metals' metabolism, as described in Section II.6, by spreadsheet routines.
The scope of FLUX is limited, i.e. it only addresses specific substance flows and stocks in a rather mechanical sense. Market and consumer behaviour are not included. No information is provided or required regarding costs or economic impacts. Within these limitations, FLUX permits evaluation of the present metals management regime and estimation of the potential effectiveness of policy measures and packages in terms of the resultant metals flows and stocks and in terms of the developed sustainability indicators. No comments can be made regarding policy instruments, implementation, feasibility, etc. and no insight is provided into the possible occurrence problem-shifting phenomena when metals emissions are replaced by emissions of other substances.
FLUX can be applied also as a dynamic model. However, this application yields less robust outcomes, partly because of the different requirements of dynamic models and partly because the possibilities for modelling the waste and emission flows from stocks are limited. The main drawback is that there is no way to model discarding-from-stock as a delayed inflow into stock. In Section IV.3 we have tested the influence of this on results, and it appears to be considerable. The dynamic mode of FLUX should therefore be considered as a first attempt at dynamic substance flow modelling and the conclusions regarded merely as indicative.

Dynabox, described in Section II.5, is a model calculating substance flows and stocks related to environmental compartments based on physico-chemical substance characteristics. It also contains a risk assessment module, which translates these flows and stocks into environmental concentrations and human intake and compares these to policy standards or no-effect levels. Dynabox has been used for the following purposes:

- to calculate steady-state environmental flows and stocks, 'the ultimate consequence of the present regime'
- to calculate the ecosystem and human health risk ratios occurring in the steady state
- to compare alternative regimes with the present regime on a steady-state basis.

Dynabox is a general environmental multimedia model. This implies that the concentrations calculated are average concentrations for an entire region. In reality concentrations may vary widely across locations, depending on local circumstances in terms of environmental characteristics and emissions. Dynabox is oblivious of any considerations regarding economic flows and management. The emissions are entered into the model as exogenous inputs with any uncertainties in these emissions being ignored. Within these limitations Dynabox can be used to evaluate certain management regimes in terms of environmental health risk ratios.
Dynabox also has a dynamic mode, which has been used to specify time paths and to calculate transgression periods. There are no specific additional drawbacks in this case.

The combination of FLUX and Dynabox thus covers the economic and environmental flows and stocks of the metals completely. There is one major drawback, however: economy-environment cycles are not adequately treated, since the connection is only one way. The emissions to emerge from FLUX are entered into Dynabox, which then calculates the environmental flows and concentrations. Extractions from the environment are not modelled in Dynabox. Thus, the harvesting of agricultural products or the extraction of contaminated sludge from sediments are not visible and these flows cannot be 'extracted' from Dynabox to be entered in FLUX. This means that FLUX and Dynabox are not totally compatible. We have no insight into the influence of this on calculation results.

In addition to the FLUX-Dynabox combination, two other models have been used to help evaluate management options: a dynamic M-P chain model for window frames and a dynamic soil balance model for agricultural soils.

The dynamic Materials-Product chain model, described in Section III.2.5, is an economic market model combined with a substance flow model for a functional system, in this case determined by the processes involved in the production and use of window frames. Included in the model are both technical and economic processes, thus linking substance flows to economic mechanisms. Its contribution to the evaluation of management options for copper and zinc in Section IV has been:

- a description of the flows of zinc, PVC and galvanised steel connected with the production and use of window frames
- an assessment of the influence of a tax on zinc on the flows of zinc and PVC over time.

Although the M-P chain system constitutes a small part of overall zinc flows and M-P chain modelling is by definition limited to small systems, it provides a valuable complement to SFA modelling, much as described in Section II.7. In the first place, it includes economic mechanisms, thereby adding causality to the system. Such modelling of the driving forces behind the substance flows is entirely lacking in SFA models. In the second place, M-P chain models can include more materials and therefore cover a greater proportion of environmental interventions. It thus allows any shifting of problems to be observed, which also is outside the scope of an SFA approach.

The DSCB model, described in Section II.4 and applied in III.3, includes a local specification of soil processes depending on agricultural management on the one hand, and soil properties on the other. It can be classified as a specific brand of SFA model with internalised soil processes with agricultural management superimposed. It can be used to assess local variations in flows and concentrations in soils. Although the DSCB system is limited in scope as well as spatially and generalisation is difficult, DSCB modelling constitutes a valuable addition to Dynabox in the analysis of management options. On the one hand it provides a check on the Dynabox outcomes. Although the USES core of Dynabox is in general use, it has been validated only in part, while the DSCB model is based on measurements of environmental flows and concentrations in real situations. On the other hand, it gives an impression of the ranges in outcome that may be expected in a real situation, additional to the average outcomes of Dynabox. On a national scale, input reduction is clearly required to prevent large-scale accumulation. On a smaller scale, however, depletion may occur for specific cases, as was shown in III.3.2. In that case, further input reduction is not recommendable at all. Despite the fact that generalisation is difficult at the moment, the 'two-way street' approach of Dynabox analysis for discovering general trends and D[SC]B modelling for assessing local variations in flows and soil contents is quite powerful.

The sustainability indices defined in II.4 enable screening and allow different agro-ecosystems to be compared without having to know all processes in detail and thus permit proper assessment of different farming systems. This exercise may yield the following information:
1) which heavy metal is likely to cause the greatest violation of standards (policy prioritisation of metals);
2) which environmental compartment is threatened most in the steady state and which compartment is likely to face problems first;
2) which experimental data assessment should have priority with a view to reducing uncertainty and optimising predictions;
3) which approaches to preventing violation of standards are feasible and most effective.

An interesting field of application is the incorporation of sustainability indices in a generic data set such as a Geographical Information System (GIS). If the parameters in a GIS could be combined in such a way that the input and output rate parameters were known, this could be very promising for large-scale (e.g. whole-region) assessments. However, a well-defined large-scale dynamic analysis is impossible at the present time owing to lack of required data. For the moment, the top-down and bottom-up results can be used in a complementary way as part of a 'research chain' in which basic research (using refined models and basic data) and holistic research (using generic models and lumped parameters) are coupled. With the former, investigations specific to sites and farming systems can be carried out, while general trends are discovered via the latter. This 'two-way street' involves up- and down-scaling (see also Table 1).

IV.4.2 Discussion and conclusions regarding sustainable metals management

From Part III it can be concluded that, given the chosen interpretation of the concepts of sustainability and environmental quality, the current Dutch metals management regime is not sustainable. Although at present no environmental quality standards are transgressed, they will be in the future, owing to releases from presently growing societal metals stocks. From the analysis in Part IV the conclusions can be added that in the case of metals, the time spans involved in transgressing such standards are sometimes extremely long. This can be regarded as positive - at least we have ample time to come up with solutions. It can also be seen as a dangerous aspect, related to the non-degradability of the metals: despite the fact that the metals' chains are already quite well managed, with a high efficiency and a high recycling rate, in the long run they still cause problems.
From the analysis in Section IV.2 it can be concluded that the main problems arise from:

- non-functional and functional agricultural trace applications (fertiliser and fodder additions)
- involuntary losses from bulk applications in the built environment
- leaching from discarded applications at landfill sites.

A sustainable management regime should therefore address these applications. Transboundary inflow via rivers is an additional problem, especially for aquatic ecosystems.

A 'moderate' scenario comprising relatively undisruptive measures does not appear to conform to sustainability. Although for zinc there is a significant reduction of risk ratios, for copper the problem remains, especially for aquatic ecosystems. In order to conform to environmental quality standards, a 'stringent' scenario, including serious changes in society, is required. Agricultural practice should change significantly, building materials should be substituted on a large scale, and waste management techniques should be adopted that are at present non-existent, not only in the Netherlands but also in upstream Rhine and Meuse countries. The environmental impacts resulting from such changes are not addressed here, but might also be considerable. The same can be said regarding the costs and economic impacts of such changes.

Developing a solution for these problems is therefore an example of an unavoidable clash between the economy and the environment. Most of the emissions considered might be reduced adequately by technical means. For these it is merely a question of being prepared to pay a certain amount of money. Cleaning up the water is another matter, however, and implies changing the whole nature of food production as well as that of residential building and waste processing. It is then a question of choice: do we choose to solve the problems to the degree laid down by the various environmental quality standards, or do we choose to keep basic societal practices more or less as they are at present? In the short term, the possibilities for reshaping the management regime in a more sustainable direction are limited. In the long run, however, more drastic changes are indicated, since the environmental problems related to metals seem to be long-winded indeed.

Part V: Summary, Conclusions and Recommendations

Jeroen C.J.M. van den Bergh, Jos Boelens, Mathijs N. Bouman, Jeroen B. Guinée, Reinout Heijungs, Gjalt Huppes, Patricia P.A.A.H. Kandelaars, Theo M. Lexmond, Simon W. Moolenaar, Lauran van Oers, Xander Olsthoorn, Helias A. Udo de Haes, Evert Verkuijlen, Ester van der Voet

Contents:

In Part V, we return to the research problem, the actual and potential environmental consequences of the production and use of heavy metals, and the connected research questions. After summarising the main results, we discuss the developed methods and models in V.1, their weaknesses and strengths and their applicability, stand-alone and in combination with others. The nature of the metals problem and its underlying causes are also discussed as well as the possibilities for establishing a more sustainable metals management regime. In V.2 the most important conclusions are presented and in V.3 some recommendations for future research as well as for policies on heavy metals are formulated.

V.1 Summary of results

V.1.1 Research questions

In Part I of this book the research questions of the Metals programme were introduced. Research question 1 refers to obtaining an overview of the flows and stocks of the four selected heavy metals in the Netherlands. To establish the linkages between the economy and the environment in a quantitative manner such an overview is not sufficient. From the perspective of managing these metals, the relations between the flows and stocks, and between the flows and economic or environmental variables, are important. These relations are the subject of research question 2: how can these flows and stocks be modelled? A number of models have been developed in the course of the research programme, as described in Part II. The application of these models to describe and analyse (parts of) the problem of heavy metals is treated in Part III. The core models developed and used in the Metals programme are *metabolic models*, describing (aspects of) the metals' metabolism in the Netherlands, i.e. the physical relations between the flows and stocks of the four heavy metals in the Dutch economy and environment. In addition, models have been developed that address the economic *driving forces* behind these flows and stocks. Establishing a link between the driving forces and the flows and stocks has been attempted successfully at the micro-level. Another important aspect concerns the *scale level* on which the models are operational. This contributes to the validation of the models, to the analysis of uncertainty and also to the identification of mechanisms of spatial problem-shifting. In Section V.1.2 the developed models and methods are discussed.

In Section V.1.3 we return to the research questions regarding a suitable metals management regime, as presented in Part I. What have the combined efforts and the developed methods and models contributed to answering these questions, what insights have been gained and what issues have remained unresolved? We discuss these aspects in two clusters. The first cluster refers to research questions 1 and 3: can we trace the fate of the metals that enter the Dutch economy but are no longer emitted to the environment, and are there any environmental risks involved? This has been investigated by linking an inventory of the metals' metabolism to an environmental fate and risk assessment model. The second cluster refers to the present metals management regime, i.e. research questions 4 and 5: is present management sustainable, and if not, what is the nature of a sustainable management regime? Section V.1.3 contains the main results and a discussion on the subject of sustainable management.

V.1.2 Developed methods and models

Metabolic models
The basic concept of models describing the flows and stocks in society is that of *industrial metabolism*, as described in Section I.2.2. This concept argues the analogy between processes in the economy and the environment on a material level: as

organisms process materials and energy in order to create biomass, so materials and energy are processed in the economic system to create 'technomass'. Models that describe this processing of materials we call *metabolic models*. A number of these have been developed in the course of the Metals programme:

- FLUX, a model of substance flows and stocks in the Dutch economy as a whole
- Dynabox, a model of substance flows and stocks in the Dutch environment as a whole
- DSCB, a model of substance flows and stocks in specific agricultural soils
- a spreadsheet model of substance flows and stocks in the Dutch housing sector, with a dynamic extension for copper in drinking water installations.

These models have in common that they are based on the mass-balance principle derived from the law of mass conservation and describe flows and stocks of metals as a result of physical interactions only. Economic and behavioural driving forces behind these flows are externalities. This limits the explanatory and forecasting power of such models, but allows for simultaneous inclusion of large systems with a great number of flows and stocks because of the simplicity of the relations.

For the overall picture of the metals' flows and stocks in the Dutch economy and environment a combination of FLUX, Dynabox and a number of specifically defined sustainability indicators were used. FLUX has been applied to the flows and stocks in the economy, Dynabox for the environmental flows and stocks. The sustainability indicators refer to flows and stocks both in the economy and in the environment and are used to streamline the evaluation of different management options.

FLUX (see Part II.2 for a description) has been used for the following purposes:

- to inventory flows and stocks in, out and through the economy for the base year 1990
- to calculate steady-state societal flows and stocks resulting from the present management regime
- to compare alternative management regimes with the present regime on a steady-state basis.

Within the limitations of metabolic models, FLUX permits evaluation of the present metals management regime and allows the potential effectiveness of measures and measure packages to be assessed in terms of the resultant flows and stocks of the metals. Thus, the strong point of FLUX is its identification of possible routes to improve metals management. However, no comments can be made regarding policy instruments, implementation, costs, feasibility etc., and no insight is provided into the possible occurrence of problem-shifting phenomena when metal emissions are replaced by emissions of other substances.

FLUX can also be applied as a dynamic model, providing time-series estimates of flows and stocks in future years. However, this application yields less robust outcomes, partly because of the different requirements of dynamic models and partly because modelling the waste and emission flows from stocks is a new area of research. This is described in more detail in Section IV.3. We have tested various possibilities for dynamic modelling, each leading to very different outcomes. Below we consider this in greater depth. The main conclusion here is that the dynamic mode of FLUX should be considered as a first attempt at dynamic substance flow modelling and that the conclusions should therefore be regarded as indicative only.

Dynabox, described in Section II.5, is a model calculating substance flows and stocks in, out and through environmental compartments based on physico-chemical substance characteristics. It also contains a risk assessment module, which translates these flows and stocks into environmental concentrations and human intake. It is based on the generally used model USES (Uniform System for the Evaluation of Substances), which has been adapted for metals. Dynabox has been used for the following purposes:

- to calculate the steady-state environmental flows and stocks resulting from the present management regime
- to calculate the ecosystem and human health risk ratios occurring in the steady state
- to compare alternative management regimes with the present regime on a steady-state basis.

Dynabox is a general environmental multimedia model. This implies that the concentrations calculated are average concentrations for the whole region under study. In reality, concentrations may vary widely across locations, depending on local circumstances regarding environmental characteristics and emissions. Dynabox is oblivious of any considerations regarding the flows in the economy and their management. The emissions are entered into the model as exogenous inputs and any uncertainties in these emissions are disregarded by Dynabox. Within these limitations, Dynabox can be used to evaluate management regimes in terms of environmental and human health risk ratios.

Dynabox has also been extended with a dynamic mode, which has been used to specify time paths and calculate the periods involved in transgressing certain concentration standards. No specific additional drawbacks are attached in this case.

In Section II.6 a number of sustainability indicators are defined, which have been used to evaluate the present metals management regime and compare this regime with alternatives, as described in Part IV. Indicators are defined in three groups, connected to the aforementioned research questions:

1. Indicators for the fate of the metals entering the economic system. Such indicators are strongly connected to the categories of flows within the defined system: export, emission, landfill, accumulation. In most cases they can be extracted directly from the overview of flows and stocks generated with FLUX. They require only a clear and precise definition of the system boundaries.
2. Indicators for the environmental risks attached to a given management regime. These indicators refer to the environmental flows and especially stocks as calculated with Dynabox. Accumulations in environmental stocks are translated into concentrations, which are then compared with Maximum Permissible Concentration (MPC) values to indicate ecosystem health risks. Human intake via the various routes is also calculated and likewise compared to Acceptable Daily Intake (ADI) values to indicate human health risks.
3. Indicators for the management of the metals. These once again refer to the flows and stocks in the economy and indicate the possibilities for improving metals management. Most of these indicators are ratios: technical process efficiency, recycling rate, fraction of trace applications. Such indicators refer to the feasibility

of a stepped-up 'closing of cycles' policy, to the appropriateness of a waste management policy, to a shift of focus from production processes to use and waste management processes, to a volume policy rather than an efficiency policy, etc. etc., as potential options for improvement.

These indicators are thus translations of the information generated with FLUX and Dynabox, aimed at a arriving at structured evaluation of the metals' metabolism. In line with FLUX and Dynabox, they are limited to the physical aspects of the metals' flows and stocks. Thus, they offer no information on policy instruments to bring about the changes indicated, nor on any economic impacts of such changes. Moreover, they provide no indication of problem-shifting to other materials or substances. The indicators are defined per substance chain. Aggregation over the group of metals is therefore not possible; it is possible, however, to compare the four chains of metals on aspects of fate, environmental risks and appropriate options for improving management.

The combination of FLUX, Dynabox and indicators thus covers the metals' flows and stocks in the Dutch economy and environment completely. The theoretical and practical integration of substance flow modelling with on the one hand risk assessment and on the other management options as two aspects of metabolism is new and, in our opinion, one of the main achievements of the Metals programme. It has thus become possible to link the inflow of materials into the economy systematically and quantitatively to environmental risks, via societal and environmental metabolism. Much more work remains to be done in this area, for example on the compatibility of the different models and the development of additional indicators, as mentioned in the various sections. However, the general framework appears to be valuable and practicable, of course within the above-mentioned limitations inherent in the metabolism approach.

The last of the developed metabolic models, the DSCB (Dynamic Soil Composition Balance) model, operates at a more detailed level. In this dynamic model, the influence of agricultural management practices on the metals' concentrations in soils is quantified, with due allowance for soil characteristics. It therefore also combines societal and environmental metabolic variables in one model, focusing on agricultural soils at the farm level only. The conclusions that can be drawn from this model are therefore quite different: such conclusions are valid only for the specific combination of soil, crop and management being modelled and may not be extended to the general Dutch situation. However, on this detailed level the model is much more sophisticated and subtle than the FLUX-Dynabox combination, thus increasing the explanatory and especially the forecasting power of the model.

The analysis of stocks in the Dutch housing sector has been valuable in obtaining greater insight into the dynamics of the four metals. There are basically two approaches to modelling the generation of emissions and waste flows from stocks. The first approach we call the *leaching* model. This is the one used in FLUX and is quite straightforward: the generation of waste and emissions from a specific societal stock is modelled as a constant fraction of the size of that stock at that particular point in time. The second approach, which we call the *delay* model, starts from the assumption that

the output from stocks in society - the generation of waste and emissions - is determined by the past input into and the residence time in those stocks. The outflow from a certain stock in a certain year thus equals the inflow into that stock of a number of years earlier. To build such a model, data are required on the present size as well as the historical build-up of societal stocks of substances, or – alternatively – lengthy time series on the stocks' inflows and outflows.
The leaching model is simpler and therefore easier to apply, but proves not to be appropriate in all cases. A comparison, described in Section IV.3.2 for the case of copper water pipes, shows that the results differ considerably, not only for the long term but also - and sometimes especially - for the short term. Further development of a delay model is therefore recommended.

Modelling driving forces
The metabolic models described above do not consider the driving forces behind the metals' flows: the demand for certain materials, products and services and the associated market mechanisms. These driving forces are the subject of economics. As a general framework for combining the modelling of materials flows with that of economic driving forces, the concept of a material-product (M-P) chain was adopted. In Section II.3 this concept is described in detail. An M-P chain is defined as a system of linked flows of materials and products supporting the provision of a certain service. The M-P chain consists of extraction, material production, production of products, recycling, reuse and treatment of final waste. Chain management can be linked to this as an overall policy strategy that explicitly considers sequential linkages between various activities in terms of both economic and physical mechanisms. In M-P chain analysis the linkages of particular activities allows the indirect effects of policies to be traced, both on environmental and on economic variables.

Based on the concept of an M-P chain various types of analysis can be performed:
1. One of the basic M-P chain models is a *static optimisation model* that describes an 'environmental manager' optimising costs under a particular set of physical and technological restrictions. The choices that need to be made concern: the nature and respective volumes of new and recycled materials in the production process, the split between new and reused products in meeting demand for the service, and materials recycling percentages. With this type of model the optimal recycling and reuse rates and the optimal input mix of materials in the production function can be assessed. The model was applied to rain gutters, where a distinction was made between zinc and PVC gutters. Alternative policies were studied, such as a product charge, recycling standards and subsidies. The results showed that the current situation is not optimal, in the sense that the net costs of demand are not minimum.
2. Another basic model of an M-P chain has a *general equilibrium* format, i.e. it permits study of market processes in relation to M-P chains. The results show that the externalities caused by extraction and generation of harmful waste can be optimised by imposing a tax on new (raw) materials. A change in certain variables causes a shift of the optimal tax structure: instead of taxing the extraction activity it may be better to tax the waste treatment activity of the M-P chain. This implies that the whole M-P chain needs to be considered when analysing optimal policy packages. It is also important to recognise that regulatory material policies aimed at reducing the use of specific materials generate revenues. A general equilibrium

model approach can take account of the redistribution of these revenues to taxpayers by lowering the tax on labour, i.e. by shifting the tax burden from labour to use of materials (i.e. environmental tax reform).

3. A dynamic analysis of M-P chains was performed in which a life-cycle assessment (LCA) was combined with an economic analysis. This *dynamic descriptive model* includes the environmental impacts of the M-P chain, accumulation and delayed effects. The model was applied to the case of window frames. Two policy packages were examined, aimed at reducing the depletion of raw materials and at reducing water pollution. The results showed that imposing a mix of policy instruments in order to prevent depletion may reduce raw materials depletion by over 50% in 20 years compared with the scenario with no additional policy. A multi-criteria analysis was performed to assess the effects of single instruments on the environmental indicators. In this way the best instrument can be identified. A dynamic analysis is especially appropriate when durable products are involved.

These are examples of physical and economic realities integrated in one model. Such examples support the analysis especially when seeking solutions for the heavy metals problem, by adding economic to physical mechanisms. The usefulness of M-P chain analysis can be further increased, for example by:

- linking different types of behavioural models to materials flows and calculating the impacts of taxes and other economic instruments
- expanding the possibilities for substitution in the existing models
- building cost-benefit models that include all types of strategies (recycling, dematerialisation, immobilisation, landfill), enabling comparison of such strategies.

Integration of metabolic and driving force models

The M-P chain model is one of the few economic models that can be combined with the modelling of materials flows. It is limited to small parts of the total picture of the metals' flows and stocks, however, and therefore cannot be directly combined with the developed metabolic models. In Part II.7 two possible routes are indicated for combining the two types of models: (1) sequential use and (2) integration of the two modelling principles in one all-encompassing model.

One further development in the direction of sequential use might be to design a procedure for subsequent use of metabolic and economic models in such a manner that the strong points of each are combined while sidestepping the limitations. Keeping to the example of heavy metals, we can imagine a heuristic approach as follows:

- use SFA models to identify the metal flows, distinguish the problematical flows, select the main flows to regulate and try out the problem-solving potential of several technically defined management options
- then, use LCA to evaluate the emerging alternatives (products, materials or product processes) with respect to shifts to other environmental problems
- next, use economic models to model the markets connected with the selected flows-to-regulate and evaluate the various possible instruments on their environmental as well as economic consequences

- finally, enter the results for the most promising options from the economic model back into the SFA model to identify unexpected problem-shifting to other parts of the substance chain.

In this way, all models have their proper sequential place without transgressing their natural bounds, while at the same time supporting evaluation much more strongly together than alone. Theoretically, this may be the easiest way to proceed. In practice, this would imply close cooperation between disciplines, which may not be easy but could certainly be worthwhile.

Quite a different direction of thinking is to attempt an integration of the modelling principles of the three types of models, in order to develop new models that have all the advantages and none of the drawbacks. M-P chain models, adding mass-balance equations to a modelling of markets, can be regarded as such, but are operational only at the micro-level. On the macro-level models exist that add one or two markets to a large input-output-like physical structure. Such modelling might be extended to include more markets. Economic and physical mechanisms are thus combined into one model. The main danger of progressing in this direction is falling into the trap of trying to design 'the ultimate model' which can do everything at the same time. In practice it may well be that in the process of integration some of the specific assets of the specialist models are lost.

Which of these two routes is the most useful, and how to proceed on each of them, cannot be decided at present. For the moment it would seem useful to try both. It may depend on the specific question that needs answering. It may even be a matter of taste. In any case this seems to be a field for research still wide open for the future.

Scale levels and problem-shifting

Scale levels are an important aspect when evaluating management options. A high spatial scale permits comprehensive evaluation, but details are lost. Fine-scale evaluations show that such details may be very important, but are blind to larger-scale impacts such as a shifting of problems to other areas or other time periods. The spatial scale level primarily selected in the Metals programme is the national level: the management of the metals in, out and through the geographic area of the Netherlands as a whole. In two respects we went further:

- by determining the 'export of pollution' from the Netherlands we have put Dutch metals management in a global perspective
- by zooming in on various aspects we were able to add more detail to the sometimes crude average analysis.

The export of pollution is one of the sustainability indicators defined in Section II.6. It specifies the sum total of metals emissions associated with the consumption of metal applications by the Dutch population, wherever and whenever these emissions take place in the world. These emissions are then compared with the emissions actually occurring within the Dutch borders, enabling conclusions to be drawn regarding the shifting of problems to other countries.

More detail has been added in various ways. In the first place, detail is added by application of M-P chain analysis to a number of metal applications as described above. In the second place, detail is added by taking a closer look at the Dutch building sector, where the largest flows and stocks of the metals occur. The developments in the housing sector and the historical build-up of several of the larger stocks are described, which has opened up possibilities for a more sophisticated, dynamic analysis. In the third place, a change of scale level is introduced by DSCB modelling at the local level. DSCB modelling addresses the agricultural sector, where the flows of metals are relatively small but where the contribution to environmental risks is high, owing to the direct impact on exposure routes.

Field-scale and farm-gate balances give farmers specific feedback on effective options for improved heavy-metals management. Farm-gate balances show the total contaminating potential of the agricultural management regime in force, whereas field-scale balances enable a direct link with soil-protection and other environmental criteria. Such specific analyses are required because of the great differences, for different metals, among and within farming systems. Dynamic field-scale balances enable field-specific analyses of heavy-metal accumulation, leaching and uptake and consequently identification of 'hot spots' (e.g. specific fields, crops, applications). The DSCB approach is presented as a way to calculate the mass balances of both heavy metals and principal soil constituents. This approach is of special relevance if organic soil amendments (e.g. manure and compost) and soil-bound heavy-metal flows (e.g. erosion) are involved. Crop rotation and fertiliser choice have a significant influence on the heavy-metals balance of arable farming systems. Only by long-term monitoring is it possible to measure the magnitude and the direction of changes in soil properties of potential consequence for heavy-metal availability and mobility.

Sustainability indices derived from the dynamic balance are proposed as indicators of adverse effects of current agricultural practices and to assess the effects of different management options designed to prevent quality standards being exceeded (see Sections II.4 and III.3). Moreover, they enable screening and comparison of different agro-ecosystems without having to know all the processes in detail. Indicators have been defined that provide information on the following issues:

- prioritisation among different metals, by comparing transgression of quality standards
- identification of the environmental compartment threatened most, or first
- prioritisation of data assessment with a view to reducing uncertainty and optimising predictions
- identification of farm management options to avoid violating standards.

DSCB modelling adds information to the average results from Dynabox: it shows the variability and the deviations from this average in both directions. It can also be regarded as a test for Dynabox, since the DSCB model is, contrary to Dynabox, validated with actual field measurements. As mentioned in Section III.1, an attempt has been made to scale up the DSCB results for cadmium to the level of the total Dutch agricultural soil mass. This yielded figures of the same order of magnitude, adding credibility to the Dynabox results.

Despite the aforementioned efforts, we must conclude that the developed models have not yet addressed aspects of spatial scale in a sufficiently consistent way. Further research is required on how to include locational aspects and how to strike a practical yet sound balance between a 'mainlines' approach and a sophisticated but highly specialised detailed approach. Combining DSCB models with GIS modelling, as mentioned in IV.4, seems to be an interesting possibility.

Apart from spatial scales, there are also time scales. In the case of heavy metals, the relevant time period to consider is large due to these substances' non-degradability. The metals accounts cover the situation in one year only. The calculations with the dynamic models cover a period of roughly a century. From a planning perspective this is already an extensive period. Meaningful forecasts of the development of flows and stocks over the next century certainly constitute valuable information for environmental policy. From the above it has become clear, hopefully, that the developed dynamic models are as yet not sufficiently robust for this purpose. The dynamics of societal stocks, in particular, are as yet insufficiently known or understood. In this direction, a field of research is still wide open.

Even dynamic models that do provide adequate results can scarcely be trusted a century into the future, for too many uncertainties emerge after such a long period of time. However, the build-up of environmental stocks extends for many more centuries, as shown by the DSCB model in Section III.3. This is where steady-state modelling offers additional value. Steady-state models calculate the equilibrium situation that ultimately results from maintaining a certain management regime, no matter how long it takes to reach. Comparing the calculations using the dynamic mode of Dynabox for the year 2100 with the steady state, we conclude that after 100 years the concentration build-up has barely started. Without offering information on the time period involved, the steady state teaches us that developments do not stop after a century. The steady state may in no event be interpreted as a forecast or prediction. Nevertheless, it offers a basis for comparison between different management regimes that is more robust, although even less realistic, than the 2100 forecasts with dynamic models. The steady state can be regarded as an exaggerated representation of the present management regime, magnifying its inherent characteristics, as opposed to dynamic modelling, which evokes expectations of reliability and accurateness. When such expectations cannot be adequately met, it may be more appropriate to adopt a 'sketchy' approach and use steady-state modelling.

V.1.3 The flows and stocks of heavy metals in the Netherlands

The fate of the mined metals and the related environmental risks

With the SFA model FLUX the flows and stocks of the four selected heavy metals (copper, zinc, cadmium and lead) in, out and through the Dutch economy have been mapped for the year 1990. Thus we obtained a picture of the metabolism of the four metals. From this exercise, the main flows and stocks of the metals have been identified, including the main sources and sinks (see Section III.1). It can be concluded that:

- all emissions are accounted for, so there are no 'hidden emissions'
- the emissions are indeed minor compared to the inflow into the Dutch economy

- the amount of landfilled metals is considerably higher than the amount of emitted metals, but still only accounts for 20% (cadmium) - 50% (lead) of the inflow
- the main sink in 1990 is the Dutch economy itself: the stocks of various applications of these metals appears to be building up quite rapidly.

In addition, application of the pollution export indicator shows that for none of the metals is Dutch society exporting its problems. Owing to the large refinery and production sector, the Dutch pollution footprint is significantly smaller than the sum total of emissions occurring within the Dutch borders. In fact, the Netherlands harbours pollution related to consumption elsewhere.

The next question is whether there are any environmental risks involved, now or in the future. At present ecological and human risk levels are not, on average, transgressed. In the future, however, the growing stocks may lead to a new increase in the generation of waste and emissions. Calculations with FLUX show that the steady-state emissions are indeed considerably larger than the 1990 emissions, as a result of the increased societal stocks. These emissions were entered into the environmental fate model Dynabox, which showed that steady-state concentrations in soil and surface water exceed Dutch environmental quality standards, sometimes severely, for three of the four metals: copper, lead and zinc. The same is true for human health standards. For lead, the problems can be expected to disappear owing to the penetration of lead-free gasoline since 1990. Surprisingly, cadmium concentrations appear to remain within the standards even in the steady state. For copper and zinc, however, there is no sign of these problems being resolved in the future. Steady-state calculations provide no information on the time span involved in transgression of environmental quality standards. However, calculations with the dynamic versions of FLUX and Dynabox show that such transgression periods may be extensive (decades to centuries).

The emission reduction currently taking place is thus at the expense of increased stock-building in the societal system. In the absence of mitigating measures and policies, this will most probably lead to a new rise in emissions in the future and a concurrent transgression of environmental quality standards. By our definition in Section I.2, the present patterns of use of copper and zinc, particularly, therefore cannot be regarded as sustainable.

Causes and origins of environmental risks

Taking a closer look at the causes of the (future) problems, several general conclusions can be drawn:

- For aquatic ecosystems, the main source at present for all four metals is what enters the Dutch territory via the Rhine and Meuse. Any abatement activities in the Netherlands are therefore futile until such time as the upstream countries also change their metals management regime.
- The technical efficiency of the Dutch processes of production, use and waste management, as well as the recycling rate, are at present quite high for all four metals. This implies that solutions must be found in other directions, especially substitution.
- Trace applications, including non-functional occurrences of the metals as contaminants, are easily overlooked because of their small size and economic

invisibility. Nevertheless, their contribution to overall emissions is large. Solutions should therefore specifically include these trace applications.

- We have found several instances of 'closed loop accumulation': the slow process of increasing amounts of trace metals occurring in economy-environment cycles. The most important closed loop is related to agriculture. Owing to the cyclic nature of these processes, the initially small input of metals with fertiliser and as a fodder additive is continually boosted, so that manure becomes the major source of metals input in agricultural soils. To what extent this closed loop accumulation will actually occur is unclear, since it may be counteracted by certain mechanisms (see IV.3.4), which are not included in the model. However, these mechanisms will result in other environmental flows, such as an increased leaching, which are also potentially harmful.
- In the steady-state situation, there is a very noticeable increase in the leaching of metals from landfill sites, constituting the greatest contribution to the aquatic ecosystem. This is due to the nature of the applied models: the models are based on the mass-conservation principle. The steady-state situation implies that what flows in has to come out again, and therefore the leakages from sites equal the inflows. Although this is a modelling peculiarity and has nothing to do with any realistic expectations, it focuses the mind on the need for a sound waste treatment policy aimed at immobilising non-degradable toxic substances such as heavy metals. Another possibility is to regard landfill sites as mines offering potential for future metals extraction, thus also alleviating possible problems of resource depletion. This notion is certainly worth exploring further in both theoretical and practical studies.

These observations relate to all four metals. In addition, though, a variety of key sources can be identified for each specific metal. For copper, water pipes can be mentioned as an important source: corrosion from these pipes makes an important contribution to emissions to surface water via the sewer system. For zinc, corrosion from applications in the built environment, such as gutters, roofs and fences, is a major source of emissions. In 1990, the main problematical application for lead was still was as a petrol additive. Since then this source has become less important and is expected to disappear completely, as mentioned above. The main cadmium-related problems arise from non-functional trace applications in zinc and phosphate, in which cadmium occurs as a contaminant.

Directions for sustainable metals management

In Section IV a number of options for managing the metals problem are considered. As it was concluded from analysis of the problem that zinc and copper show the largest increase in future emissions, we concentrated on those two metals. We used a number of models, both dynamic and steady-state, to assess abatement options. A reference scenario, defined as the continuation of the present management regime by freezing the technical coefficients, was compared with two alternatives:

- a 'moderate' scenario comprising a number of reasonable measures in the agricultural and housing sectors, with a view to assessing the problem-solving power of readily feasible options

- a 'stringent scenario' proceeding from the moderate scenario but adding such measures as are required to meet environmental quality standards, now and in the long term.

The moderate scenario comprises technical emission reduction, termination of certain harmful applications (pesticides), more efficient use of copper in pipe systems, substitution of zinc by stainless steel or plastics in building applications and a reduction of agricultural applications by 10 - 30%. Calculations with FLUX and Dynabox show that this package reduces environmental risks somewhat but in neither case sufficiently, although the time period involved in transgressing the standards becomes even longer. More stringent measures are therefore required to conform to sustainability. An iterative process leads to definition of a 'stringent' scenario, which includes serious changes in society. If environmental quality standards are to be met in the long run, agricultural practice must change significantly, building materials must be substituted on a large scale and waste management techniques aimed at immobilisation must be adopted that are at present non-existent, not only in the Netherlands but also in upstream Rhine and Meuse countries.

When considering changes of such a magnitude, the limitations of the analysis as performed here become very restrictive. In the first place, there is no insight into the trade-off among environmental impacts as a result of substitutions and other changes in society. The production, use and waste management of other materials also leads to emissions, which may in turn cause environmental standards to be transgressed. In the second place, there is no insight into the economic and societal impacts of such changes, which may be considerable but also strongly depend on the time period involved in realising them. Without a proper understanding of these issues, analysis is incomplete and results should therefore be interpreted carefully. Further research in this area is required to put the problem of heavy metals as well as its possible solutions into perspective.

Perspectives for the future
The main, robust conclusion from the above is that, in order to conform to sustainability as defined in Part I.2, major changes are required in the metals management regime, since the environmental problems related to metals seem to be long-winded indeed. These changes go beyond specific metals applications, to include a basic shift in building materials, a severe reduction of livestock production and the development of waste-processing techniques that do not exist at present. It is then a question of choice: do we choose to solve the problems to the degree laid down by the various environmental quality standards, or do we choose to keep basic societal practices more or less as they are at present? This question has various aspects to consider.

In the first place, the feasibility of such changes is an issue. In the short term, the possibilities for developing management in a sustainable direction are limited. In the long run there is greater scope but this would imply careful planning. Governments may not be inclined to do so, since metals are only part of the environmental problem, and not even an urgent part according to public opinion. Therefore priorities may well be elsewhere. Still, environmental standards even for low-priority substances are an important part of environmental policy and abolishing them would certainly be a drastic step. The fact that the problem has lost its newness, and therefore public attention, does not mean that it should not be further addressed.

In the second place, an entirely different view may be taken regarding the issue of sustainability. The stringent scenario of Part IV is a typical example of a 'strong sustainability' view: the environmental quality standards are absolutes and must be reached by any means, the only latitude available to government is to find the least-cost way of doing so. When a 'weak sustainability' view is adopted, implying that natural capital is at least partially exchangeable with man-made capital, the standards become less absolute and there is far more scope for allocation of funds based on a weighing-up of priorities. A scenario based on cost-benefit analysis may look very different from one based on cost-effectiveness within the constraints of environmental standards. This would be an interesting road to explore.

A third issue concerns the time-frame of the problem. The time periods involved in the transgression of environmental standards are quite long. On the one hand, this lessens the urgency: a problem that occurs only after centuries does not have to be addressed immediately. This conforms to notions of discounting, common in economic theory: the farther ahead the problem is, the less it counts when we evaluate it now. On the other hand, such an approach conflicts with the future-generations aspect of sustainability: we should not impose our problems on future generations, even very future ones, which we clearly do in the case of heavy metals. We therefore conclude, rather, that the long transition periods do allow for slow but persistent tightening up of policy, eventually resulting in the major changes required.

V.2 Conclusions

V.2.1 Developed models and methods

1. Within the Metals programme, two types of models have been developed to address the problem of heavy metals: (1) metabolic models, describing the flows and stocks of heavy metals in the Dutch economy and environment, and (2) behavioural models, describing driving forces behind some of the metals' applications.

2. The developed metabolic models have three modes: accounting, static/steady-state and dynamic. The accounting mode can be used to organise data into one comprehensive overview of flows and stocks for a specific year. Static/steady-state modelling can be used to relate environmental problem flows to their origins in the economy and to compare different management options. The dynamic mode can be used for forecasting flows and stocks in future years.

3. Metabolic substance flow models derive their strength from the rigorous application of the mass-balance principle. This enables problem-shifting to be detected in both time and space. On the other hand, they have a limited scope: problem-shifting to other substance chains cannot be detected, nor do they include market and consumer behaviour.

4. By linking metabolic substance flow models to generic risk assessment models it is possible to specify, in a quantitative manner, the whole chain of events from the inflow of metals into the economic system, through production, use and waste management, via emissions into the environment, the environmental fate and the various intake routes, finally to a measure for human and ecosystem health risks.

5. The choice of scale level and system demarcation is of major influence on the outcome of modelling applications. A detailed scale level adds information that is invisible on the national level, but certain mechanisms operating at a higher scale level are then beyond the model's scope. Analysis benefits from applying models at different scale levels, preferably covering the entire range from global to local.

6. The developed behavioural models are economic models extended with mass balances. They address a number of linked markets of metals' applications (Materials-Product or M-P Chains). Static M-P chain models can be used for cost optimisation of the chain under certain restrictions. Equilibrium M-P chain models can be used to assess the influence of financial instruments on markets. A dynamic M-P chain model can be used to trace developments in materials flows and prices over time.

7. Behavioural models permit substitution and therefore also identification of problem-shifting to other substance chains. The main limitation of such models is that only a small fraction of all relevant markets can be included.

8. The various metabolic and behavioural models may sometimes lead to seemingly contradictory results. On closer observation, these apparent contradictions originate from the different scopes of the models: they represent different aspects of reality, and therefore yield additional rather than contradictory results. The analysis benefits from applying models with different scopes.

9. The development of sustainability indicators additional to the models has proven very useful, on the one hand for focusing the analysis and on the other for comparing different management options in uniform metrics.

V.2.2 Heavy metals in the Netherlands

10. Over the past few decades, emissions of copper, zinc, lead and cadmium to air and water have decreased considerably. The inflow into the Dutch economy has remained at the same level, however. Instead of being emitted, these metals now accumulate in the Dutch economy in a number of stocks of products, and in a number of closed loops connected in particular with agriculture.

11. The Netherlands does not engage in problem-shifting to other countries: the emissions of cadmium, zinc and copper taking place worldwide on behalf of Dutch consumption are smaller than the emissions taking place within the Netherlands.

12. The accumulation does not cause risk levels to be transgressed at present. However, if the present management regime continues to be pursued emissions will rise again in the future, owing to the increased stocks, causing transgression of human and ecosystem health standards, especially for copper and zinc. The present management regime therefore implies problem-shifting to the future.

13. The time periods involved in the transgression of environmental standards are lengthy: from several decades to many centuries. On the one hand, this implies that problems related to heavy metals are less urgent than others. On the other hand, the fact must also be considered that the response to mitigating measures also will be very slow, because of the long residence time of metals in both economic and environmental stocks.

14. The emissions increase will take place despite the policy measures already taken, which have resulted in quite high technical process efficiencies as well as recycling rate.

15. The main problems will arise from three sources: (1) trace applications in agriculture, such as additions of copper and zinc to fodder and contaminants in fertiliser, (2) bulk applications in the built environment, such as water pipes, gutters and roofs, and (3) leakage from landfill sites.

16. At the local level, inputs and accumulation of heavy metals in agricultural soils are highly influenced by the combination of crop, soil and agricultural practice. Wide differences can be observed compared to the Dutch average, in both directions.

17. A 'moderate' scenario, comprising relatively feasible measures, reduces environmental health risks but, from a strong sustainability perspective, insufficiently so.

18. In order to conform to policy standards, more stringent measures are required, including a basic shift in building materials, a severe reduction of livestock production and the development of waste processing techniques aimed at immobilisation that do not exist at present. However, there is ample time to bring about such changes.

19. The extensive changes in society required to conform to policy standards cannot be brought about by taxes on metals, since material costs only constitute a limited part of the product price even for the larger-scale functional applications of the metals.

20. On the farm level, a trade-off appears to exist between cadmium, on the one hand, and copper and zinc, on the other: measures combating the input of cadmium into agricultural soils raise the inputs of copper and zinc, and vice versa.

V.3 Recommendations

V.3.1 Developed models and methods

1. The models contain various areas of uncertainty. Some uncertainties are related to the availability and quality of basic data. Heavy metals are relatively well-investigated and documented. The main problems are connected with stock data, which are virtually non-existent at present. Another type of uncertainty is introduced when translating basic data on substance flows into modelling parameters. Consistent investigation of the influence of such uncertainties on the results of the model calculations is recommended.

2. Dynamic modelling is very useful for gaining insight into the impacts of certain management regimes over time. In an SFA context such modelling is only in its earliest stages of development. The modelling approach taken here represents only one of possible approaches. It has been demonstrated that other types of dynamic models lead to different outcomes. This implies that the results presented here should not be taken absolutely but merely as indicative of direction. Further development of dynamic substance flow models is strongly recommended in order to obtain insight into the mechanisms behind the observed growth and decline of societal and environmental stocks.

3. The use of formalised indicators has proven valuable. In the course of analysis the need for several additional indicators emerged: (1) in view of the importance of losses from landfill sites, an indicator for immobilisation, (2) in view of the relative importance of trace applications, an indicator for dissipation of substances within the economic system, and (3) an indicator for the now completely out-of-sight depletion of metals stocks.

4. In the analysis it has been concluded that consideration of different spatial scale levels is relevant, owing to the fact that different mechanisms occur at different scale levels. In the Metals programme we included the local and the national level. Other relevant levels are: the European level (trade and policy), the level of the Rhine and Meuse basins (catchment areas) and the global level (both trade and global environmental processes). Consistent investigation of the problem-causing as well as the potentially problem-solving mechanisms is recommended on the various scale levels.

5. As observed above, it has proven very useful to apply models of differing scope to one and the same problem, in order to cover as many sides of the story as possible. In the analysis presented here this was performed in a rather haphazard manner. A more systematic investigation into the different possibilities for proceeding in this area is recommended. We envisage sensible possibilities in two directions: (1) expanding combinations of driving forces and materials flows in one model, and

(2) designing procedures to use models of differing scope in parallel and/or in series.

V.3.2 Heavy metals in the Netherlands

6. The closed loop accumulation that was detected by the metabolic models is based on the law of mass conservation. There are other underlying assumptions beside this law, especially the assumption that plant uptake is proportional to concentration in the soil. There are indications that this is not a valid assumption in all cases. More research is required into the mechanisms underlying the uptake of metals by crops to sustain the conclusion of closed loop accumulation.

7. An important source of future emissions appears to be losses from landfill sites. This conclusion is based solely on the law of mass conservation as it applies in the steady state. More insight is required into the actual processes taking place at landfill sites and how these affect metals, to be able to formulate an adequate waste management policy for heavy metals.

8. One of the major limitations of the metabolic models is their inability to spot the influence of certain changes on other substance chains. For a balanced evaluation of management options, however, this is vital information. M-P chain models have the ability to include substitutes for specific applications but do not cover large-scale societal changes either. Additional methods and models should be developed to include possible problem-shifting on a larger scale in the analysis.

9. Another hiatus in the developed approach concerns an evaluation of the economic impacts of the implied societal changes. The second- and third-order economic impacts of major changes in agriculture or a new style of building are left completely out of the picture at present. These impacts may be substantial but may also be positive. How to include such aspects is a question to which there is currently no answer and therefore one that needs addressing urgently.

10. In the Metals programme, a strict approach has been taken towards the concept of sustainability, which can be characterised as a 'strong sustainability' approach. Environmental standards with regard to metals are treated as constraints for economic development. This implies that there is no scope for weighing up the environmental benefits against the costs required to meet these constraints. A 'weak sustainability' approach, allowing for trade-off between natural and human-made capital, would be more flexible in that respect and would enable an analysis in terms of how best to spend one's money. It would be a valuable extension to compare such a weak sustainability approach to the strong sustainability approach taken as the starting point for the analysis of the Metals programme.

Appendix 1 Glossary

term	meaning
General concepts and terms	
biogeochemical cycles	Definition: Natural global cycles of nutrients as driven by biological, chemical and geological processes. Elaboration: Such cycles are of crucial importance for the functioning of the biosphere. By human addition of nutrients biogeochemical cycles may be enlarged, by addition of micro-pollutants they may be disturbed.
ecosystem	Definition: A community of organisms and their environment, including also the interactions between the organisms and between the organisms and their environment. Elaboration: Ecosystems can be defined at different scale levels ("from the whole globe down to a single drop of water"). Ecosystem processes play an important role in keeping the global cycles going.
indicator	Definition: A measure of a specific policy goal or objective. Elaboration: In this study, both economic and physical indicators are used. Economic indicators refer to economic goals and values, for example the net costs, the total demand, allocation variables, government revenue of levies, average cost per product, changes in employment and trade, real output change of production sectors and household groups. Physical indicators refer, directly or indirectly, to environmental goals and values. Examples are: emission of specific substances, depletion of specific resources, biodiversity loss, but also stockbuilding in the economic subsystem, recycling rate, technical efficiency of processes.
industrial ecology	Definition: The science that studies the processes in the physical economy in analogy to the study of the processes in the (physical) environment. Elaboration: There is no generally accepted definition of industrial ecology. Some consider "industrial ecology" equivalent to "industrial metabolism", others to "industrial ecosystems". However most seem to agree that industrial ecology should describe a research area or a branch of science. Also there is agreement on the central issue: the analogy

term	meaning
	between the biosphere and the technosphere, and on the central subject of research: the physical economy.
industrial ecosystem	Definition: A community of industries designed to optimize processes according to the notion of industrial metabolism. Elaboration: The crucial issue is the combination of processes related to different industries. This can be combined services, for example of energy generation and waste water treatment. It can be finding opportunities to use waste streams, for example by generating energy from waste flows. It can also be by locating related industries on one site, in order to reduce transport.
industrial metabolism	Definition: The analogy between the use of materials and energy in the technosphere to create "technomass" and the use of materials and energy in the biosphere to create biomass by metabolic processes. Elaboration: In the biosphere, cycles are generally closed and process efficiency is optimal. By taking lessons from the biosphere, economic processes can be improved and attuned to minimize losses to the environment. Especially the combination of processes - one process's waste is another's resource - is crucial.
mass balance principle	Definition: The mass balance principle states that in a closed system materials cannot be lost, only altered (derived from the first law of thermodynamics). Elaboration: For an open system the mass balance principle implies that all materials that go into a system either leave the system or accumulate in the system.
Modelling concepts and terms	
accounting (= bookkeeping)	Definition: Keeping track of flows and stocks by registering them. Elaboration: Accounting or bookkeeping enables policy makers to spot trends, and to monitor the effects of certain changes including policy measures afterwards.
accumulation / desaccumulation	Definition: Increase / decrease in one or more of the available stocks of the system. Elaboration: Accumulation in the economic subsystem occurs in products that are produced but not discarded in the same year. Desaccumulation occurs when discarding exceeds inflow,

term	meaning
	e.g. in cases of terminated applications. Accumulation in the environment mainly occurs in soils and sediments, in the groundwater, and in standing stocks of biota. Desaccumulation in the environment occurs when soils and sites with a formerly high input may leak stockpiled substances.
allocation	Definition: The attribution of a quantity to various objects, for example, of a material to various products, or of labour to various industrial sectors. Elaboration: Allocation can be performed from different points of view, resulting in different attributions.
closed system	Definition: A system with supposedly no relations to a possible environment. Elaboration: In terms of substance flow studies, closed systems hardly occur. The earth as a whole may be viewed as a closed system for matter, although even this is a simplification.
dynamic model	Definition: A model specifying the relations between entities including time as a modelling parameter. Elaboration: With a dynamic model, scannings of the future may be performed for specific substance management regimes, indicating the development of flows and stocks over time for a specific time period.
economic subsystem (=technosphere, anthroposphere, economic domain, societal subsystem)	Definition: Those nodes in the system referring to anthropogenic processes regarding production, use and discarding of materials / goods / products / services. Elaboration: Some processes occur actually on the border of the economic and environmental subsystems, such as agriculture and landfill. A choice must be made which is always debatable. In this study, the choice was made to consider agricultural soil as part of the environment, while landfill sites are located within the economic subsystem.
emission (= leak, = leakage)	Definition: Flow from the economic subsystem to the environmental subsystem. Elaboration: The question which flows can be classified as emissions is determined by the demarcation between the economic and environmental subsystems (see above). In this study, fertiliser application is an emission. The landfilling of final waste is a flow within the economic subsystem, but leaching flows from landfill sites to the groundwater are emissions.

term	meaning
environmental subsystem (=biosphere, ecosphere, environmental domain)	Definition: Those nodes in the system referring to natural processes. Elaboration: Natural processes such as degradation in soils, chemical reactions in the atmosphere, erosion etc., which take place without (major) human control, can be classified as environmental or geological. The environmental subsystem may or may not include geological processes. In the latter case, a third subsystem is required: the geological subsystem.
extraction	Definition: Flow from the environmental subsystem to the economic subsystem. Elaboration: The question which flows are extractions is determined by the demarcation between the economic and environmental subsystems (see above). In this study, harvesting is classified as an extraction, and so is dredging and landfilling of harbour sediments.
flow	Definition: The transport of a specific amount of a substance / material / product from one node to another within a specific period of time. Elaboration: All three categories – substance, material, product - exist. When drawing up a substance account, a translation is required to express flows of materials and products in terms of the amount of the substance they represent. For the time period, a year is chosen. Flows occur within the economic subsystem, within the environmental subsystem and between the subsystems.
formation / degradation	Definition: The coming into existence, resp. ceasing to exist, of substances by chemical formation and degradation. Elaboration: Formation and degradation may occur through processes in both subsystems. In the case of metals, neither formation nor degradation takes place.
functional flow descriptors	Definition: Functional flow descriptors describe substance flow carriers in functional terms. Elaboration: Functional flow descriptors may refer to origins (e.g. cowsmilk), function (e.g. lubricating oil), or destination (e.g. chemical waste to-be-processed).
geological subsystem (=geosphere, lithosphere, geological domain)	Definition: Those nodes in a substance account referring to geological processes. Elaboration: Usually, the geological processes occur on such a time scale that investigating them in a substance flow study is hardly relevant. The quantification of geological processes

term	meaning
	may thus be limited to the exchange with the economic and environmental subsystems by erosion, volcanic activity, mining and immobilization.
immobile stocks	Definition: Stocks that under the given environmental conditions do not interact with the flows and stocks in the economic and environmental subsystems, nor are influenced by economic or environmental processes. Elaboration: Geological stocks can in the present context be classified as immobile. This category may also encompass stocks in the economic subsystem, if sufficiently isolated from both economic and environmental processes.
import/inflow and export/outflow	Definition: Flows entering the region from, resp. leaving the region to, neighbouring regions. Elaboration: In the economic subsystem, this concerns the import and export of the substance in raw materials, goods and waste products; in the environmental subsystem, it is the transboundary in- and outflow via air and surface water.
Materials-Product chains (M-P chains)	Definition: A set of linked material and product flows connected to a certain service or application. Elaboration: This refers to a network of economic activities between extraction of resources and waste treatment, connected via flows of materials and products. An M-P chain is a chain of processes in economic (functional and monetary) and physical dimensions. This combination allows one to include recycling of materials, reuse of products and substitution in an M-P chain.
model	Definition: Simplified simulation or representation of (aspects of) the real world. Elaboration: A substance flow model belongs to the category of systems analysis models. It describes a region's substance management regime (the present one or a modified one) in terms of the exchange of the substance between the nodes of the system representing the region. Three types of substance flow models can be distinguished: accounting systems, static models and dynamic models.
node	Definition: Point of redistribution or transformation of substance flows. Elaboration: A separate element in the substance account or model, representing a (group of) proces(ses). To every node, one or more (mutually related) inflows and outflows of

term	meaning
	substances can be linked. Within a node accumulation can take place in a connected stock.
open system	Definition: System with explicit relations to a specified surroundings. Elaboration: In terms of substance flow studies, the relations of the system to the surroundings consist of import / transboundary inflow and export / transboundary outflow.
physical flow descriptors	Definition: Physical flow descriptors describe substance flow carriers in physical terms. Elaboration: Physical flow descriptors can be: elements; molecules; agglomerates of elements and compounds (e.g. iron ore; eutrophicating nitrogen compounds; plastics); mass (kg); energy (e.g. higher/lower combustion value); form (ball/sphere; tile; aquous solution; powder; etc.)
process	Definition: Unit transforming inputs into outputs. Elaboration: The transformation can be intentional or non-intentional. There are economic processes, environmental processes and geological processes.
production factor	Definition: Input of an economic process. Elaboration: The three production factors usually distinguished in economics are: capital, labour and environment.
recovery/erosion (=mobilization) and immobilization	Definition: Flows from immobile stocks to either the economic or the environmental subsystem, resp. flows from the economic or environmental subsystem to immobile stocks. Elaboration: The becoming available, resp. becoming unavailable, of substances because of extraction from or addition to the system's immobile stocks. This may be caused by environmental processes (erosion, covering) or economic processes (mining, waste disposal).
sink	Definition: Node with an inflow from other nodes within the system, but with no outflow to other nodes within the system. Elaboration: sinks can be classified as export, transboundary outflow, immobilisation by economic or environmental processes, and chemical degradation by economic or environmental processes.
source	Definition: Node with an outflow to other nodes within the system, but with no inflow from other nodes within the system. Elaboration: sources can be classified as import, transboundary inflow, mobilisation by mining or erosion, and chemical for-

term	meaning
	ming by economic or environmental processes of a substance.
static model	Definition: Model specifying the relations between entities excluding time as a modelling parameter. Elaboration: a static model may be used to describe the present situation, but also for determining the steady state situation belonging to a certain substance management regime.
steady state model (= comparative static model)	Definition: Model specifying the equilibrium situation that will be reached by keeping modelling parameters constant, which can be used to compare scenarios on their long-term consequences. Elaboration: In the case of substance models, the steady state is the overview of flows and stocks that arises from calculating the equilibrium situation of a specific substance management regime. Any difference with the overview for the present situation is due to an initial dis-equilibrium in stocks, by accumulation or desaccumulation, which has disappeared in the steady state. Some regard steady state models as a form of static modelling, others as a specific form of dynamic modelling.
stock	Definition: Amount of a substance / material / product / resource stored within a node. Elaboration: There are stocks in the economic subsystem (materials and products), stocks in the environmental subsystem (amounts of a substance stored in environmental compartments) and stocks in the geological subsystem (amounts of a substance or resource stored in the earth's crust).
substance flow diagram	Definition: A schematic summary of a substance account. Elaboration: In past studies by CML, the term Substance Flow Diagram was used to indicate a specific type of summary, i.e. a summary wherein categories of flows are distinguished. Other types of summary, for example a summary based on categories of nodes, could also be referred to as substance flow diagrams.
substance management regime	Definition: The technical (modelling) characteristics of the substance's pathways in, out and through the economic subsystem as a result of intentional and non-intentional management of economic processes. Elaboration: In terms of a substance flow model, the substance regime is described by the matrix of distribution coefficients.
substance/ materials	Definition: An aggregated overview of flows and stocks of a particular substance or material in a particular region for a

term	meaning
account	particular time period, as obtained by bookkeeping. Elaboration: Such aggregated overviews can also be obtained by (static or dynamic) modelling. In that case the overview is simply referred to as "the overview of flows and stocks".
subsystem	Definition: Ordered set of connected processes within a larger system. Elaboration: Apart from the economic, environmental and geological subsystems, other subsystems may be distinguished, referring for example to the phases in the life cycle of the substance, or to otherwise coherent groups of nodes.
system	Definition: Ordered set of connected processes. Elaboration: In terms of substance flow studies, the system is an ordered set of nodes, connected by substance flows, and possibly containing one or more stocks.
value flow descriptors	Definition: Value flow descriptors describe substance flow carriers in terms of their economic value. Elaboration: The economic value of flows generally is expressed in monetary terms.

Policy terms and concepts

base scenario / business-as-usual scenario	Definition: Scenario including autonomous socio-economic developments and already accepted policy assumptions, but no additional policy measures or exogenous changes. Elaboration: This type of scenario is not included in this study. A distinction is made between a base / business-as-usual scenario including socio-economic developments and a reference scenario which does not necessarily include such developments.
environmental sustainability (ecological sustainability)	Definition: Environmental sustainability is guaranteed when environmental interventions are kept within the limits of the environmental carrying capacity. Elaboration: In this study, environmental sustainability is interpreted as: keeping within the maximum level of metals emissions that still is compatible with environmental quality (see below).
environmental quality	Definition: Quality of the environment in terms of human and ecosystem health and well-being . Elaboration: In this study, ecosystem and human health is

term	meaning
	assumed to be protected when environmental policy standards regarding metals for soil, sediments and surface water and for the daily intake of humans are met.
integrated life cycle management (= integrated chain management)	Definition: Management of functional chains from-cradle-to-grave. Elaboration: "the integrated chain" may refer to substance chains, product chains, materials chains, or mass chains. Integrated life cycle management aims at minimizing extractions from the environment, closing economic cycles as far as possible, and keeping residual emissions and waste flows within acceptable limits.
policy scenario	Definition: Scenario including possible policy measures, generally but not necessarily on top of autonomous socio-economic developments. Elaboration: In this study, policy scenarios are defined without autonomous developments. Moreover, the scenarios are operationalized in terms of technical measures and not in terms of the policy instruments required for implementation.
recycling	Definition: Collecting and dismantling, granulating or transforming of disposed products into their constituent parts to be used for new applications. Elaboration: Recycling is distinguished from reuse (see below). However sometimes the term recycling is used as encompassing both reuse and recycling *sensu stricto*.
reference scenario	Definition: Scenario used as a basis for comparison with policy scenarios. Elaboration: In this study, the reference scenario does neither contain socio-economic developments nor additional policy measures, but is defined as the steady state situation of the 1990 management regime of the metals.
reuse	Definition: the collecting, cleaning and using again, without dismantling, of disposed products for the same purpose. Elaboration: re-use is distinguished from recycling (see above).
risk	Definition: The probablity of undesirable environmental impacts due to environmental interventions. Elaboration: This term is adopted in the Dutch environmental policy and applied for disasters and emissions of substances. The risk is calculated from the probability of actual adverse impacts on human and ecosystem health, often combined with

term	meaning
	the probability of the occurring of the event. In ecological risk assessment, referring to continuous emissions of toxic substances, the risk is calculated from the estimated fate and the estimated effects of the substance. Ecological risk assessment was used in this study.
scenario	Definition: Scenarios are argumented and logically structured future visions, which may include estimated autonomous developments as well as policy induced developments, often lacking a statement about the probability of realisation. Elaboration: In this study, scenarios are built out of (policy) measures reducing metal emissions on top of the management regime of 1990. For reasons of comparability and simplicity, they are evaluated by steady state modelling. Autonomous socio-economic developments have been excluded in this study.
substitution	Definition: Replacement of a material / process / product by another, functionally equivalent, material / process / product. Elaboration: A distinction is made between direct substitution (between materials) and indirect substitution (between materials and non-materials). When formulating scenarios for heavy metals, direct substitution of metal-containing products by metal-free ones is one of the possible policy measures. Since the analysis is limited to the metals themselves, the scenarios cannot be evaluated regarding the substitution impacts, i.e. the change in extractions and emissions outside the realm of the metals.
sustainable development	Definition: Societal development that meets the needs of the present generation without compromising the ability of future generations to meet their own needs. Generally, three aspects are distinguished: the environmental, the economic and the socio-psychological aspect. Elaboration: In the case of a substance management regime, sustainable development is interpreted as such a management regime that allows for environmental sustainability, now and in the future.

Methods and tools of analysis

(Environmental) Life	Definition: LCA is a tool for assessing the environmental

term	meaning
Cycle Assessment (LCA)	impacts of a from-cradle-to-grave product system required for a particular unit of function. Elaboration: The term "product system" is taken to mean the product throughout its entire life-cycle, from cradle to grave, in terms of all the economic processes involved. A product system can be regarded as a specific type of M-P chain. LCA can be used as an optimising tool but also to compare alternative product systems on their environmental consequences.
Equilibrium model analysis (GEA and PEA)	Definition: An analysis of the economic consequences of certain policy measures, using a model in which agents optimize behaviour, budgets are balanced and all markets are cleared. Elaboration: General equilibrium analysis (GEA) includes all markets, while partial equilibrium analysis (PEA) considers one or a few markets while taking the others as exogenous. An applied general equilibrium (AGE) model is applied to, for example, a country or a region for which the effects of a certain policy on economic groups, including the government, are examined. PEA is applied to systems on the micro-level, for example to M-P chains.
Input-output analysis (IOA)	Definition: The analysis of the mutual exchanges of goods and services between different industrial sectors and towards final users. Elaboration: The result of IOA is an input-output table which describes these exchanges either in monetary units or in physical ones. For purposes of adding environmental parameters, IOA tables have been extended with emission factors; alternatively waste management sectors and the environment have been introduced as sectors in the table to make the exchanges with waste and the environment possible.
Material Flow Accounting (Material Flow Analysis, MFA)	Definition: The accounting / analysis of flows of materials in, out and through an economy-environment system. Elaboration: Materials may refer to mass in general, to specific materials (PVC, paper, steel), and to substances (elements and compounds). The system usually is regionally bounded, but in some cases system boundaries are defined on a functional basis. Sometimes, specific attention is paid to environmental flows. In other cases the analysis or accounting is restricted to

term	meaning
	flows in the economic subsystem, including extractions from and emissions to the environment.
Materials-Product chain analysis (MPCA)	Definition: The analysis of a Materials-Product chain (see above). Elaboration: M-P chain analysis can take various shapes, depending on the purpose of the analysis. In general, MPCA is used for the study of optimization, market equilibrium, market processes, production functions, substitution at different levels by explicit modelling of economic processes, and endogenous behaviour of agents. It can be used to optimize the environmental impact of M-P chains, or to achieve a certain accepted environmental impact at least social costs.
Substance Flow Analysis (SFA)	Definition: The analysis of a substance's pathways in, out and through a region for a certain time period, including all economic and environmental processes involved. Elaboration: SFA is a specific type of MFA. It aims at providing the relevant information for a region's overall management strategy regarding single substances or coherent groups of substances.

ENVIRONMENT & POLICY

1. Dutch Committee for Long-Term Environmental Policy: *The Environment: Towards a Sustainable Future.* 1994 ISBN 0-7923-2655-5; Pb 0-7923-2656-3
2. O. Kuik, P. Peters and N. Schrijver (eds.): *Joint Implementation to Curb Climate Change.* Legal and Economic Aspects. 1994 ISBN 0-7923-2825-6
3. C.J. Jepma (ed.): *The Feasibility of Joint Implementation.* 1995 ISBN 0-7923-3426-4
4. F.J. Dietz, H.R.J. Vollebergh and J.L. de Vries (eds.): *Environment, Incentives and the Common Market.* 1995 ISBN 0-7923-3602-X
5. J.F.Th. Schoute, P.A. Finke, F.R. Veeneklaas and H.P. Wolfert (eds.): *Scenario Studies for the Rural Environment.* 1995 ISBN 0-7923-3748-4
6. R.E. Munn, J.W.M. la Rivière and N. van Lookeren Campagne: *Policy Making in an Era of Global Environmental Change.* 1996 ISBN 0-7923-3872-3
7. F. Oosterhuis, F. Rubik and G. Scholl: *Product Policy in Europe: New Environmental Perspectives.* 1996 ISBN 0-7923-4078-7
8. J. Gupta: *The Climate Change Convention and Developing Countries: From Conflict to Consensus?* 1997 ISBN 0-7923-4577-0
9. M. Rolén, H. Sjöberg and U. Svedin (eds.): *International Governance on Environmental Issues.* 1997 ISBN 0-7923-4701-3
10. M.A. Ridley: *Lowering the Cost of Emission Reduction: Joint Implementation in the Framework Convention on Climate Change.* 1998 ISBN 0-7923-4914-8
11. G.J.I. Schrama (ed.): *Drinking Water Supply and Agricultural Pollution.* Preventive Action by the Water Supply Sector in the European Union and the United States. 1998 ISBN 0-7923-5104-5
12. P. Glasbergen: *Co-operative Environmental Governance: Public-Private Agreements as a Policy Strategy.* 1998 ISBN 0-7923-5148-7; Pb 0-7923-5149-5
13. P. Vellinga, F. Berkhout and J. Gupta (eds.): *Managing a Material World.* Perspectives in Industrial Ecology. 1998 ISBN 0-7923-5153-3; Pb 0-7923-5206-8
14. F.H.J.M. Coenen, D. Huitema and L.J. O'Toole, Jr. (eds.): *Participation and the Quality of Environmental Decision Making.* 1998 ISBN 0-7923-5264-5
15. D.M. Pugh and J.V. Tarazona (eds.): *Regulation for Chemical Safety in Europe: Analysis, Comment and Criticism.* 1998 ISBN 0-7923-5269-6
16. W. Østreng (ed.): *National Security and International Environmental Cooperation in the Arctic – the Case of the Northern Sea Route.* 1999 ISBN 0-7923-5528-8
17. S.V. Meijerink: *Conflict and Cooperation on the Scheldt River Basin.* A Case Study of Decision Making on International Scheldt Issues between 1967 and 1997. 1999 ISBN 0-7923-5650-0
18. M.A. Mohamed Salih: *Environmental Politics and Liberation in Contemporary Africa.* 1999 ISBN 0-7923-5650-0
19. C.J. Jepma and W. van der Gaast (eds.): *On the Compatibility of Flexible Instruments.* 1999 ISBN 0-7923-5728-0
20. M. Andersson: *Change and Continuity in Poland's Environmental Policy.* 1999 ISBN 0-7923-6051-6

ENVIRONMENT & POLICY

21. W. Kägi: *Economics of Climate Change: The Contribution of Forestry Projects*. 2000
ISBN 0-7923-6103-2
22. E. van der Voet, J.B. Guinée and H.A.U. de Haes (eds.): *Heavy Metals: A Problem Solved?* Methods and Models to Evaluate Policy Strategies for Heavy Metals. 2000
ISBN 0-7923-6192-X

KLUWER ACADEMIC PUBLISHERS – DORDRECHT / BOSTON / LONDON

MIX
Papier aus verantwortungsvollen Quellen
Paper from responsible sources
FSC® C105338

If you have any concerns about our products,
you can contact us on
ProductSafety@springernature.com

In case Publisher is established outside the EU,
the EU authorized representative is:
Springer Nature Customer Service Center GmbH
Europaplatz 3, 69115 Heidelberg, Germany

Printed by Libri Plureos GmbH
in Hamburg, Germany